THE LIFE
OF
SIR J. J. THOMSON

J.J. Thomson in 1933

THE LIFE
OF
SIR J. J. THOMSON
O.M.

*Sometime Master of Trinity College
Cambridge*

by

LORD RAYLEIGH

CAMBRIDGE
AT THE UNIVERSITY PRESS
1942

CAMBRIDGE UNIVERSITY PRESS
Cambridge, New York, Melbourne, Madrid, Cape Town,
Singapore, São Paulo, Delhi, Tokyo, Mexico City

Cambridge University Press
The Edinburgh Building, Cambridge CB2 8RU, UK

Published in the United States of America by Cambridge University Press, New York

www.cambridge.org
Information on this title: www.cambridge.org/9781107655423

© Cambridge University Press 1942

First published 1942
First paperback edition 2011

A catalogue record for this publication is available from the British Library

ISBN 978-1-107-65542-3 Paperback

CONTENTS

LIST OF ILLUSTRATIONS

PREFACE

THE LIFE OF J.J. THOMSON was a full and vigorous one, and could be approached from various points of view. His own *Recollections and Reflections* (1936) must always have a first place with those who wish to understand his personality and achievements, and it must be admitted that he has himself skimmed much of the cream off what there is to be told. When it was suggested to me by his family that I might attempt a biography, I was inclined to take the view that the subject was almost exhausted. However, on further consideration, it seemed that there was room for a somewhat different treatment of his life. Some points which were prominent to outside observers were naturally omitted from his own book on grounds of modesty. Again, the account of his most important investigations, though admirable, is somewhat in the textbook style, and hardly gives much insight into the perplexities and technical difficulties which had to be overcome. Nor does it show in sufficient detail the complicated interactions between his own work and the work of his pupils, many of whom were men of power and originality, and started independent investigations which converged in a remarkable way on the final results attained by the School under his guidance. The present writer was fortunate enough to have observed much of this development at first hand, and should therefore be in a reasonably favourable position for relating it. The *Recollections and Reflections* has been consulted for facts and dates, but apart from this limited use, my account has been written independently.

If any reader is inclined to complain that the book is too scientific, I can only plead that I have tried to give due emphasis to the human side. A biographer cannot properly make the important events of his subject's life other than they were in fact, but I have put them in as easy a form as I could. No formulae have been used, in view of the protest of a well-known literary man, that he could not even skip a formula! A simple account of J. J.'s chief researches without this limitation will be found in his *Recollections and Reflections.*

I must acknowledge the generous help I have received from many of those who were associated with J. J. Thomson at different stages of his career, and who have written their recollections at length. Among these are Prof. C. G. Barkla, Dr N. R. Campbell, Sir William Dampier, Prof. A. R. Forsyth, Mr W. Craig Henderson, K.C., Prof. F. Horton, Sir Owen Richardson, Prof. D. S. Robertson, Dr G. F. C. Searle, Mr S. Skinner, Prof. L. R. Wilberforce, Prof. C. T. R. Wilson, Prof. H. A. Wilson, Mr D. A. Winstanley and Dr Alexander Wood.

Many others have helped me with smaller contributions which appear in the course of the narrative. I have not scrupled on occasion to adopt their actual words without the use of inverted commas, which rather interrupt continuity.

Above all, I have referred constantly to Lady Thomson and Miss Joan Thomson. Miss Thomson has always been ready to help me by looking up references and making enquiries on special points. I am also indebted to her for the compilation of the index.

RAYLEIGH

TERLING PLACE
CHELMSFORD

July 1942

CHAPTER I

EARLY YEARS. LIFE PREVIOUS TO APPOINTMENT AS CAVENDISH PROFESSOR

JOSEPH JOHN THOMSON came of a family who had been established in Manchester for several generations. The pedigree below gives what is known of his forbears, with a few notes about collateral relatives.

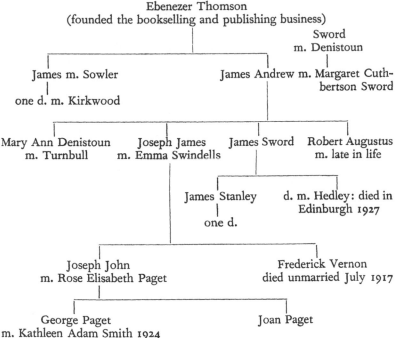

Emma Swindells was first cousin of the father of Percy Vernon (died), Charlie Vernon (died), Maud Vernon, Roland Vernon, Beatrice Vernon m. Ryan.

James Sword Thomson was a Cotton Broker in Liverpool.

Robert Augustus Thomson had no profession.

His father, Joseph James Thomson, carried on the family business of bookseller and publisher. Owing to his early death little is now remembered about him. He was of purely Scottish descent. He had one sister and two brothers, one of whom was not competent to manage his own affairs.

The bookselling business was somewhat similar to that of Quaritch in London, specialising in rare and antique books. His son later shared this taste, and was fond of picking up old books on gardening from the stalls in the market-place and elsewhere in Cambridge.

Mrs Thomson, J.J.'s mother, was the daughter of a Mr Swindells. He had one inconvenient characteristic, namely of producing a bewildering untidiness in his room, which apparently he could not help, and one of his family had to be at hand to produce some sort of order before a housemaid could start work. This peculiarity was inherited by his daughter, Emma (Mrs Thomson), in the form of inability to find her way about in either town or country. Thus, after spending a four or five weeks' holiday at Whitby with her son and daughter-in-law, she happened to be left alone after they had started for Cambridge. The luggage had gone to the station, and she had planned to walk the distance of not very many yards along a familiar road, which she liked doing. But when the time came, the fear of being 'lost', and so missing the train, compelled her to send for a cab. It is believed that other cases similar to hers have been known.

The question will naturally be asked whether there was any indication of scientific ability in the family. J.J. Thomson's own answer to this was that one of his uncles took some interest in meteorology and (perhaps) in botany.

J.J. Thomson was born at Cheetham Hill, near Manchester, on December 18th, 1856. The house is one in a terrace, so his father's business must have been carried on elsewhere.

The earliest recollections of him which are available are from Miss Gertrude Mellor, who writes:

When my sister was married to Mr Vernon who lived quite near to us, I was only nine years old, and 'Joe' Thomson, who was then eleven, was his cousin.

My first and only recollection of him at that time was at my sister's house, one afternoon, a small boy in a little grey suit and a blue silk

tie, sitting on rather a high chair, with his legs dangling—very silent—
and no doubt shy. The next time in the same place he must have been
several years older because (apropos of what I can't say) he remarked
that when he grew up, he intended to go in for 'Original Research'
and I remember my brother-in-law laughed and tapped him on his
head and said 'Don't be such a little prig, Joe'. Naturally I hadn't
the faintest idea what he meant. . . .

They lived in Plymouth Grove, a residential part of Manchester:
in as far as I can visualise it a smallish house, and I imagine in quite
a simple way.

I remember going to tea there once, when Mrs Thomson said to her
son, 'Joe, give your arm to Gertie, and bring her in', to his embarrassment I thought, for he stumbled over a footstool, and we both nearly
came to grief. It was an incident I have always remembered! He
must have been in his early teens and very shy.

I met him some years later, several times in Oxford Road near Owens
College. . .when we exchanged a few words. The last time my chief
remembrance was that he was wearing a nice brown suit, with a brown
tie (I think it had white spots) the bow of which was under one ear,
a fact of which I daresay he was quite unaware, and was possibly indifferent, and I longed to rectify. . . .

Joe Thomson was sent by his father to Owens College, Manchester, in 1871, when he was only fourteen years of age.

His father died two years later, and his mother at once received
the kindest possible offers of help from friends on whom she had
no sort of claim of a financial kind—a testimonial as to how highly
she and her husband were regarded. She removed her home
shortly afterwards to 11 Egerton Terrace, Fallowfield (now part
of Manchester), which had the advantage of being near the College.
Joe was already so absorbed in his work that he was not helpful
at home, but his brother, Frederick Vernon Thomson, who was
two years younger, was of much more use in this respect, the
more so that he could be depended on not to be absent-minded.
He eventually went in for a business career with Claflin & Co.,
calico merchants, with a large connection in the U.S.A., for the
financial means were lacking to send both sons to college, and
there seems to have been no doubt which of them was best
qualified to take advantage of it. Both he and his mother devoted much of their energies to helping with parochial church

work, and J.J.'s home life must have been passed largely in this atmosphere.

There is not much detailed information available about the home life of the Thomson family before J.J.'s marriage. Mrs Ryan (*née* Beatrice Vernon), a cousin, writes (abbreviated slightly):

My memories of the Thomsons are mainly of the annual pantomime party. All five of us with my mother were taken to a matinée. We met at the theatre, where we always sat in the middle of the front row of the dress circle, and after the performance drove out in 'growlers'* to their very comfortable Victorian house at Withington. There we had a sumptuous Lancashire high tea, and spent a lively evening, Joe being one of the liveliest of the party. He must have had a great liking for and understanding of children in those days; the parties probably took place when he was between twenty-five and thirty-five. When he was too lively his mother used to pull him up just as if he was a small boy himself, but he was quite irrepressible and her mild reproaches had little effect.

Mrs Thomson must have been a good deal older than my mother (whose hair had turned grey very early) but she still had dark hair and did not seem old to me. She wore it in a style then out of date, clusters of ringlets hanging over her ears, which I had only seen in the illustrations in Dickens' novels, and over it an arrangement of black lace and lavender ribbons. She was small, with bright dark eyes, beaming with kindness. On our departure each of us received a new florin and a large bar of chocolate cream.

Sometimes on other occasions two or three of us would tramp over to see her; we always had a very warm welcome, and a festive tea with home-made strawberry jam; and on leaving new silver and chocolate. She was obviously very proud of Joe, and used to sit and beam upon him; she showed us one of his early books, which was completely incomprehensible to her and to us!

When she heard of his engagement she was very nervous about meeting his fiancée; it would have been a great blow to her if they had not been sympathetic; when Joe brought Miss Paget† to Withington we were invited to meet her and I remember with what enthusiasm his mother spoke to us of her delight in her future daughter-in-law.

My sister used often to stay with them at the sea-side in her school holidays; Joe was about twelve years older, but an extremely enter-

* I.e. four-wheeled cabs.
† The marriage had in fact already taken place.

taining companion. He had a keen zest for the 'yellow backs' which preceded to-day's detective stories—and had a large repertoire of music-hall ditties, of which 'My Maria's a fairy queen' was a favourite. He used to sing these about the house.

We may here mention that Mrs Thomson's sons always spent their summer holidays with their mother, both before and after J.J.'s marriage. She died in 1901. J.J. chose the inscription on her monument: 'Her children rise up and call her blessed' (Proverbs xxxi. 28).

J. J. Thomson's home life had not given him any glimpse of science, and he considered that the decision to send him to Owens College was the turning-point in his life.

When there he came under the influence of Balfour Stewart, the Professor of Physics, and worked in his laboratory, which was situated in a series of attic rooms in a house in Quay Street, which had been the home of Cobden, the apostle of Free Trade. There were only about half a dozen students. The work was not much organised, and Thomson afterwards congratulated himself on having been trained under this loosely knit system. Towards the end of his life he said:

> The teaching I got at Owens College sixty-three years ago was as good as I could get anywhere if I was beginning my studies now. My first introduction to Physics was the lectures of Balfour Stewart. These were so clear that, child as I was, I could understand them.

In 1887 he wrote to Mr C. Balfour Stewart:

> Few can have been so indebted to [your father] as I was, it was he who first gave me a liking for physics and taught me most of what I know, and when I left the college I never saw him without receiving the wisest advice and the kindest encouragement, and now it is all over. Your father was the ideal man of science. No one since Faraday has ever combined such scientific genius and such deep piety.

Stewart, on his side, was accustomed to speak of Thomson as his best and most promising pupil.

His experimental work in Stewart's laboratory nearly had a tragic outcome in an explosion of a glass vessel, which injured his eyes. Mr C. Balfour Stewart writes:

My earliest recollection dates from 1874 or thereabouts when we as a family going home from church called at his mother's to enquire after J.J. who had an accident to his eyes the day before and my father felt some anxiety lest he should lose his eyesight.

However, all ended well, and his sight does not seem to have been permanently affected.

Besides Stewart, he learned much from Osborne Reynolds, Professor of Engineering, and Barker, Professor of Mathematics. The two former became, or perhaps were already, well-known names in the scientific world. Professor Barker did not achieve mathematical fame, but he was an able mathematician, and Thomson valued his teaching highly.

J.J. does not seem to have had any early training in chemistry, either at Owens College or afterwards at Cambridge; and it is possible that his later work on gases was somewhat hampered for lack of it. One observer who was favourably placed for judging did not think he had ever read a formal textbook on this subject. He got his information by asking questions; and sometimes his attitude on chemical matters seemed rather naïve. I remember when I got some unexpected results with sulphur dioxide, J.J. queried, rather unnecessarily as I thought, whether my sulphur dioxide was really sulphur dioxide at all!

Thomson's abilities made themselves felt very quickly. In June 1873 he won the Ashbury Engineering Scholarship, and the Dalton Junior Mathematical Scholarship, and in June 1874 the Dalton Senior Mathematical Scholarship and the Engineering Essay Prize.

Towards the end of his time at Owens College at the age of twenty he published in the *Proceedings of the Royal Society* a short experimental paper 'On Contact Electricity of Insulators'. It was communicated by Balfour Stewart, and made acknowledgements to him and to Dr Schuster.*

It was thus early that he came into contact with Schuster, and also with J.H. Poynting, for whom he had the greatest affection and regard, and who became a life-long friend.

---

* Afterwards Sir Arthur Schuster, F.R.S.

His father's wish had been to apprentice him to Sharpe Stewarts, locomotive manufacturers, as he mentioned at the banquet of the Institution of Mechanical Engineers held at Cambridge in 1932. Barker advised him, however, to give up this project, and to try for a scholarship at Cambridge. An Owens College contemporary writes:

I have retained a fairly clear and certainly very happy memory of him. I seem to see a rather pallid, boney youth with the air of a serious but happy student, unassuming and modest without diffidence, very approachable and friendly.

One small incident struck me at the time as typical of him, and became more interesting with his increasing eminence in after years. Going one afternoon to the Physical Laboratory for an elementary 'experiment' I found Thomson there starting on some elaborate operation—electrical with complex wiring if I remember right. During a breathing pause we fell to comparing notes on our intentions after leaving Owens. I spoke of trying for a mathematical scholarship at one of the smaller and less difficult colleges at Cambridge, and on his rejoining that he was going to sit for one at Trinity, I not then aware of his calibre, was somewhat surprised and suggested that he was flying very high. He replied, quite simply and modestly, 'I have been told I ought to get one'. I think these were the exact words; there was no 'swank', only a quiet confidence in himself which I have remembered ever since.

Thomson entered Trinity as a Minor Scholar in 1876. He remained in the technical sense a resident member of that College for the rest of his life, though only sleeping within the College walls for something less than half of it. He always said that the climate of Cambridge suited him, and that he felt well and vigorous when he was there.

As an undergraduate, he lodged at 12 Malcolm Street, a lodging-house full of poll men (i.e. men who were reading for an ordinary degree as opposed to an honours degree), with whom he kept up friendly relations. He used to say that he preferred lodgings because in College there was no one to keep up the fire, and he invariably forgot to do so when absorbed in work.

As regards his reputation with his mathematical contemporaries, Prof. A. R. Forsyth, who was a year junior, has said:

There were some . . . who even then seemed assured of future greatness and would never need to abide challenge. There was J.J. Thomson,
at that time and down to this day known as 'J.J.' in respect and general
affection. His personality stood out; we felt that he was framed in an
intellectual mould different from ours: he had our worship as completely as Alfred Lyttelton and A.G. Steel, who were our contemporaries, had secured it in the world of cricket.

In my further account of J.J. as a mathematical student I shall
mainly rely on some notes which Prof. Forsyth (who died in June
1942) kindly wrote for me, in many cases using his actual words.

As an undergraduate, Thomson attended lectures in College
by Thomas Dale, W.D. Niven and J.W.L. Glaisher, members
of the Trinity mathematical staff, and he coached with Routh,
whose private tuition was given in rooms in Peterhouse.

Niven lectured on electricity and magnetism, with special reference to Maxwell's work, at that date a comparative novelty in
the Mathematical Tripos; it was a long course, extending over
the Michaelmas and Lent terms in an academic year. J.J. attended
the course in 1877–78. Niven knew his subject well; but he did
not possess a gift of clear and connected exposition, either in
calculation or in explanation; and the result of an hour's lecture,
at any rate so far as the students whose point of view was primarily
mathematical were concerned, was a collection of brief sentences,
remarks, symbols and occasional diagrams far from comprehensible
initially. So, in the succeeding year, three of the best mathematicians—A.R. Forsyth, A.E. Steinthal and R.S. Heath—used to
adjourn to Forsyth's rooms after each lecture. With the help of
Maxwell's *Treatise* and some library books, they spent two or
three hours in making a coherent account of the substance of
Niven's lecture. In the Michaelmas term of 1879 when J.J. was
revising for his impending Tripos, he asked Forsyth for the loan
of their joint notes. They were duly lent, and found useful. It was
one of the little ironies of life that a physicist like Thomson should
have found assistance in a subject of which he was to become a
master from the joint notes of these students who were really pure
mathematicians only. It may be added that J.J. expressed warm
thanks to Niven for personal kindness and they remained intimate

J. J. Thomson in 1879

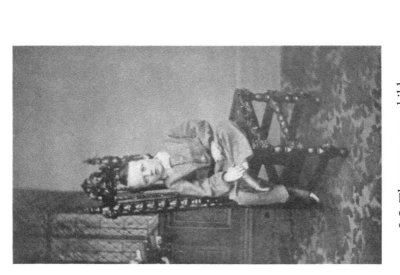

J. J. Thomson as a child

friends for the rest of Niven's life. Niven, on his side, confessed that he was rather afraid of J.J. as a pupil. Whether J.J. ever realised this does not appear.

He also attended the professorial lectures of Stokes, Adams and Cayley. Stokes' lectures on physical optics were not specially difficult and were illustrated with admirable experiments. Many went to them and enjoyed them. On the other hand, Adams and Cayley were over the heads of all but a very few, and Thomson was probably the only undergraduate of his year who attended them. Adams expressed satisfaction with his work on Jupiter's satellites which he did in connection with these lectures. Though Maxwell was then in residence as Cavendish Professor, Thomson did not come into personal contact with him.

Doubtless owing to the influence of Barker, Thomson had a working knowledge of quaternions when he went into residence at Trinity. At that time a renewed interest was being taken in Hamilton's work, largely stimulated by P.G. Tait, who was a strenuous prophet of the quaternion calculus: but the interest was a passing phase, never very active in Cambridge. One remark of Thomson's was that if he had obtained a new result by means of quaternions, he would not use it unless or until he had obtained a proof by some other method.

In the Mathematical Tripos, he came out as second Wrangler, Larmor* being the Senior. Thomson has recorded that he was satisfied with this result, the place being quite as good as he had hoped for. Though Larmor and Thomson were both taught by Routh, were friendly as undergraduates, and lived the greater part of their lives in close proximity and were interested in much the same subjects, there seems to have been little direct exchange of ideas between them at any time. There is no ground for the notion which has gained some currency that Thomson at any time asked for help from Larmor in a mathematical difficulty. It was not necessary, nor would it have been congenial to his habit of mind.

In connection with Thomson's place in the Mathematical Tripos,

* Afterwards Sir Joseph Larmor, Lucasian Professor. M.P. for the University.

it is worth putting on record that competent judges did not think him 'a good examination candidate' in a strenuous competitive mathematical examination. He did not write quickly, a valuable asset in such circumstances. He had not the flair for judging the most expedient stage at which to begin a written answer. His thoughts moved within the field of the phenomena of physics rather than in the domain of canonical mathematical ideas and formulae, resembling Maxwell's in this respect. The strain of the Tripos examination, day after day, may have troubled him, as it has troubled many; he hit upon an ingenious device for restoring his freshness by having a shampoo between the morning and afternoon papers on each day of the examination.

During his undergraduate time, J.J. did not row, nor play either cricket or football, these being the chief sporting amusements of that day: but then and always afterwards he took the keenest interest in watching cricket (particularly at Fenner's) and Rugby football. In summer he would play lawn tennis; some years passed before he began to play court tennis. In winter he would 'do' each day one or other of the regular 'grinds' in accordance with the Cambridge afternoon habit of exercise, practised by young and old alike. J.J. was a good walker, and walked on occasion as far as Royston, thirteen miles. In the evening with some select friends he often played whist, both before and after his first degree, and he became something of an adept at the game.

During Thomson's undergraduate course, there began an internal stir about mathematical study in the University, which already had decided that in the near future the whole Tripos should be changed radically. One aim was to offer freedom of choice among the more advanced subjects, the whole body corporate of which had grown too great for fair inclusion in an undergraduate's limited years of study. One or two spirits, then deemed daring, were by way of anticipating their freedom; and they proceeded to try their prentice hands at original papers, as William Thomson had done in an earlier generation. Glaisher, one of the editors of the two Cambridge mathematical periodicals, encouraged them as he always encouraged young writers. J.J. Thomson was one of these, and papers by him were printed while he was an undergraduate. One of them was 'On the resolution of the product of two sums of eight squares with the sum of eight squares' and the other was 'An extension of Arbogast's method of derivation'. They appeared in the *Messenger of Mathematics*.

These, with two other early papers, were the only ventures into pure mathematics that he ever made.

After taking his degree, Thomson began working at the Cavendish Laboratory, and also at the preparation of a thesis for the fellowship of Trinity. This was on a subject which had suggested itself to him when he was attending the lectures of Balfour Stewart at Owens College, but which had lain fallow in the meantime. It was on the transformation of energy. This thesis formed the basis of two papers in the *Philosophical Transactions of the Royal Society*, and also of a book on *Application of Dynamics to Physics and Chemistry*; also of some articles in the second edition of Watts' *Dictionary of Chemistry*, published in 1894.

Some account may here be given of the line of thought developed in these various publications. The thesis was never published in its original form. Thomson says:

A dynamical example may illustrate what the application of dynamical principles to physical problems may be expected to do, and the way in which it is likely to do it. Let us suppose that we have a number of pointers on a dial and that behind the dial the various pointers are connected by a quantity of mechanism of the nature of which we are entirely ignorant. Then, if we move one of the pointers, *A* say, it may so happen that we set another one, *B*, in motion. If we now observe how the velocity and position of *B* depend on the velocity and position of *A* we can, by the aid of dynamics, foretell the motion of *A* when the velocity and position of *B* are assigned, and we can do this even though we are ignorant of the nature of the mechanism connecting the two pointers. Or again, we may find that the motion of *B* when *A* is assigned depends to some extent on the velocity and position of a third pointer *C*: if in this case we observe the effect of the motion of *C* upon that of *A* and *B* we may deduce by dynamics the way in which the motion of *C* will be affected by the positions and velocities of the pointers *A* and *B*.

This example, given by Thomson, may be particularised further by a familiar example. The majority of people who use watches or clocks may be, and probably are, ignorant of the mechanism which connects the hour and minute hands. But they know well enough that if the position and velocity of the minute hand *A* are known, those of the hour hand *B* will follow. And conversely, when the position and velocity of the hour hand *B* are known, those of the minute hand *A* will follow. This is a typical and straightforward case. But other cases can be proposed where it would seem that the principle as above enunciated fails. Thus let *A* be carried on a sleeve rotating with strong friction on a fixed pin, and let *B* be carried on a coaxial outer sleeve moving with light friction on the first sleeve. Then, if we move *A*, *B* will copy its motion; but if we move *B*, *A* will remain at rest.

Thomson discusses the question why the general methods of dynamics should apparently fail in cases of this kind, and says:

We have hitherto made the very important restriction that the cases we considered were those where there were no resistance, frictional forces, or things such as electrical resistances, etc. If we give a wide enough meaning to the term material system, we ought to be able to deduce such forces from the dynamics of such systems by the use of the ordinary dynamical methods. Forces of this kind are assumed to be proportional to the velocities of the corresponding co-ordinates, and a steady transformation from one kind of energy into another, generally heat, is supposed to go on without any reverse transformation taking place. It can, however, I think be proved that such forces cannot be deduced from the dynamics of an ordinary system supposing the arrangements of the system to remain continuous. What I believe the equations of motion with the frictional forces inserted in the usual way give, is a result which is true on the average taken over a time which depends on the nature of the problem but is not true at any particular instant. . . ..

The methods given in the book are of too technical a nature to make it possible to attempt to give any detailed account of them here, and the results, apart from the methods, are not in themselves of popular interest. As an example, it is shown that if magnetising an iron bar alters its length, then, conversely, a change of length will affect the magnetisation, and the amount of one of these effects can be calculated from the known amount of the other. Another problem investigated is the action of an electrified atom in causing the condensation of water vapour around it. This had an important bearing on later studies in the Cavendish Laboratory, to which we shall come in due course.

At Trinity, a candidate for a fellowship has three chances, at the end of the first, second and third summers after taking his degree. It is not usual even to try at the first opportunity, partly because most candidates have not got far enough with their research to have much to put forward, and partly because electors are supposed to take the view that the junior men can very well wait. Thomson, however, succeeded at the first attempt, his great gifts having no doubt gained their proper recognition.

Shortly after this, Thomson was a candidate for a new professorship of applied mathematics at Owens College, and sent in glowing testimonials from A. Cayley, J. W. L. Glaisher, H. W. Watson, W. D. Niven and John Hopkinson. Schuster, however, who was already teaching there, was preferred.

After becoming a Fellow of Trinity, Thomson moved into College rooms (Great Court, Staircase N, No. 3). He remained in these rooms until 1885, when he moved into Nevile's Court, Staircase G, No. 2 (first floor), near the library, and he did not move again until his marriage.

These latter rooms are a fine set, and J.J. evidently took considerable pride in them. He bought some very good pieces of furniture, including a Sheraton sideboard and large cabinet and a set of Hepplewhite dining-room chairs which had belonged to the Duke of Clarence* when he was an undergraduate at Trinity. He took a considerable interest in furniture and described his acquisitions in his letters to his friend, Mrs H. F. Reid. He had photographs done of his rooms which are still extant.

In 1882 he sent in an essay which gained for him the Adams Prize, the subject set being: 'A General Investigation of the action upon each other of two closed vortices in a perfect incompressible fluid.'

The essay was published in 1883. In it the stability of inter-

* Elder brother of King George V.

locked vortex rings was investigated with great mathematical power, and it was shown that not more than seven could be linked together without breaking up into new arrangements. This study was inspired by the vortex atom theory of matter which is not now in date, but it gives the first indication of J. J. Thomson's mind working on the problems of atomic structure.

At this time he took some private pupils in mathematics, among others two men who afterwards achieved distinction—Sir Eldon Gorst, who was High Commissioner in Egypt, and Sir Austen Chamberlain, who among other high offices was Secretary of State for Foreign Affairs. Although most of his pupils were not particularly promising as mathematicians, he seems to have enjoyed the personal contacts which his teaching gave, and to have liked studying the personal capacities and idiosyncrasies of the men. He did not think it a waste of time, and was of opinion that a change in the current of his thoughts was rather advantageous than otherwise. Two years after taking his degree he was taken on to the mathematical teaching staff of Trinity on the resignation of W. D. Niven. The terms on which he was engaged required that he should give the men personal attention, and this meant rather heavy hours of teaching, which continued for about two years. At the end of this time he became a University Lecturer, lecturing on various branches of applied mathematics. In 1884 he was elected a fellow of the Royal Society, at the age of twenty-seven years. To be elected so young was not as exceptional then as it would be now, when the number of older men who have made science their career, and who have a claim to be elected in due course, is so much increased.

Immediately after taking his degree Thomson had begun to work in the Cavendish Laboratory.

Since the most active part of his life was spent as Cavendish Professor of Experimental Physics at Cambridge, it will be convenient to give here a short introductory account of the previous history of the chair and of the laboratory.*

The tradition of physical science at Cambridge had for many

---

* For a fuller account reference may be made to *A History of the Cavendish Laboratory*, 1871–1910, London, 1910.

years previous to about 1865 been somewhat antagonistic to experimental research. So far as teaching went, physics was mainly in the hands of mathematicians who had no experience in experimenting, and not much sympathy with it, at all events as a discipline for young men. There were some exceptions, such for example as Stokes and Airy, who experimented themselves, but did not make any move to teach the art to their pupils. Stokes gave beautiful experimental demonstrations in his lectures on light, but was not particularly encouraging to young men who wished to learn experimental methods from him. Other prominent mathematicians, such as Todhunter, saw no educational advantage in experiments at all, and even went to the length of preferring not to see them themselves, for fear of their ideas being upset.

Round about 1870, however, other ideas began to prevail. Laboratories were started at Oxford and at University College and King's College in London. It began to be felt at Cambridge that something must be done, but this was delayed by what J. J. Thomson afterwards called 'the usual Cambridge reason'— lack of funds. On October 10th, 1870, the seventh Duke of Devonshire—great-grandfather of the present Duke—who was Chancellor of the University and had been a good mathematician in his youth, wrote to the Vice-Chancellor offering to pay for a laboratory and apparatus. In grateful recognition of this gift the name Cavendish Laboratory was adopted from the family name of the Duke of Devonshire, and seldom has money been better expended.

The first professor was James Clerk-Maxwell, who held the office from 1871 to his death towards the end of 1879. The delays involved on the material side in getting the institution going, and Maxwell's illness, prevented the achievement of very much during his tenure. The work done by undergraduates was very unsystematic and they were to a great extent left to shift for themselves as to what they should do and how they should do it. Thomson had himself learnt under such a system at Owens College and was glad that he had done so, though he recognised of course that it could not continue when numbers became large and conditions competitive. It was in 1874 that practical work in physics was introduced into the Natural Sciences Tripos.

On Maxwell's death in 1879, the third Lord Rayleigh* was appointed to succeed him and held the post for five years. He was a landed proprietor of independent means, though somewhat embarrassed temporarily by the agricultural depression of 1879, and never intended to hold the post for any long period. However, he threw himself with energy into the work of developing the laboratory. He appointed R. T. Glazebrook† and W. N. Shaw‡ as demonstrators, and with their help organised systematic classes of practical work, mainly for students taking the Natural Sciences Tripos. In these classes the means for doing a given experiment were not duplicated, but each experiment remained set out on its own bench, and the students came round to it in turn. They had to take their chance whether they understood the theory of what they were going to do, and often learned it as they went along, with the help of the demonstrator and his manuscript instructions. These instructions were embodied in a book *Practical Physics* by Glazebrook and Shaw.

The research work done at the Cavendish in Lord Rayleigh's time was mainly on the determination of the absolute electrical units,§ which he carried to a degree of precision much higher than had been attained by earlier workers. He had adopted this line of work partly because he found that those who were already at work in the laboratory had under Maxwell's influence become deeply interested in electrical questions. It must be remembered that at this time electrical engineering was only just being brought to birth; there was, for example, no supply of current in the laboratory other than from primary batteries; and electricity as a subject of study had by no means the preponderant importance which it has since assumed. Rayleigh thought that the scheme of

* Father of the present writer.
† Afterwards Sir Richard Glazebrook, K.C.B., F.R.S., first director of the National Physical Laboratory.
‡ Now Sir Napier Shaw, F.R.S., Director of the Meteorological Office, 1905–20.
§ A simple account will be found in the *Life of Lord Rayleigh* by the present writer, London, 1924.

determining the electrical units would afford a programme in which other workers in the laboratory could co-operate, according to their capacities, and he enlisted the help of Glazebrook, Shaw, his sister-in-law Mrs Sidgwick, Arthur Schuster, who at this time came to work in the laboratory, J. M. Dodds, and J. J. Thomson.

Thomson was initiated into the problem of determining ' v ', the ratio of the electromagnetic to the electrostatic unit of electrical quantity. I do not think it would be useful or desirable to enter here into any detailed explanation of the nature of this rather technical problem. It may seem to some readers that such a question can be of very little scientific interest—of no more interest for example than the ratio of the pound to the kilogram, or the yard to the metre. These latter ratios, of course, depend on the arbitrary definition of the units in question, which are not related to one another in any natural way. It is quite otherwise with the two systems of electrical units, which are not independent, but are defined in such a way that their relation to one another is the expression of an important constant of nature. The ratio of two quantities of electricity is in one sense a pure number. For example, we can obtain the masses of silver which they can respectively deposit and the ratio of these masses is as much a pure number as the ratio of the standard pound to the standard kilogram. In another sense, however, the ratio of the electromagnetic to the electrostatic unit is not a pure number, but a velocity, because the numeric expressing this ratio depends on the units of length and time in the same way that the numeric expressing a velocity does. If we increased the centimetre, the number expressing the velocity of the earth in its orbit (e.g.) would be diminished proportionately. If we increased the second, the same number would be increased proportionately. The number expressing the ratio of the units would be affected in exactly the same way, and thus it too expresses a velocity. According to Maxwell's electromagnetic theory of light, this velocity should be the same as the velocity of light. Without detailed understanding it should be of general interest that measurements made with batteries, resistance coils, condensers, galvanometer, and so on, should be thought capable of giving the velocity of light, originally determined by observations on Jupiter's satellites, and later by terrestrial observation on light transmitted over a few miles. It was, and is, of crucial importance to prove or disprove the exact equality of these velocities. If the relation were not exactly true, we should expect, with improving technique and increased precision of measurement, to disprove it. If (as is believed to be the case) it is exactly true, all that improved technique can do is to reduce the possible margin of uncertainty: for every human measurement has limited precision: and time often shows that the errors are larger than what the sanguine hopes of the experimenter had assigned as possible.

It will be understood then that what Thomson undertook at this time was nothing radically novel. He hoped to improve on the accuracy obtained by earlier workers. Rayleigh had already designed some of the apparatus to be used, and had contemplated taking part in the work himself. It would perhaps have been better if he had done so, but, as he mentioned to me many years later, 'Thomson rather ran away with it', a natural result of energy, enthusiasm and self-reliance. Work of this kind, which is nothing if not highly accurate, is, however, full of traps and pitfalls, and perusal of the published paper suggests that the author whose experience in experiments was rather limited for so ambitious an undertaking was over sanguine that he had foreseen the possible sources of error, without applying the test of using alternative methods. He makes a remark, quoted below, which shows that the facilities available were, to say the least, the reverse of luxurious:

It may be worthy of remark that as many of the pieces of apparatus used were required for the ordinary work of the laboratory, the whole arrangement had to be taken down and put together again between each determination. This must have had the effect of getting rid of a good many accidental errors. . . .

The result of the investigation came out about 1 per cent lower than what later workers have found, and when this became apparent Thomson himself was not satisfied with it, as we shall see. He repeated the work some seven years later in collaboration with G. F. C. Searle, and they succeeded in tracing some unsuspected causes of error.

Towards the end of 1884, it became known that Lord Rayleigh was resigning the Cavendish Professorship. Sir William Thomson (later Lord Kelvin) had been approached with a view to persuading him to stand for it in 1871 when the chair was founded, and again on Maxwell's death in 1879. He was now approached for a third time, J. J. Thomson being the active mover in collecting signatures. Informed opinion at Cambridge was practically unanimous in wishing to get him. However, it proved that he was still unwilling to leave Glasgow.

Lord Rayleigh definitely resigned at the end of 1884, and the board of electors to fill up the appointment was constituted thus— the Vice-Chancellor (Dr Ferrers), Prof. Clifton, Prof. G. H. Darwin, Sir William Grove, Prof. Liveing, Prof. W. D. Niven, Prof. Stokes, Prof. James Stuart, Sir William Thomson. It is known that they found their task a difficult and invidious one, but they decided in favour of J. J. Thomson. It is believed that Niven was one of those who pressed his claims most strongly.

CHAPTER II

EARLY DAYS OF THE CAVENDISH
PROFESSORSHIP

AT the time of his appointment to the Cavendish Professorship J. J. Thomson was very young. There were few, if any, recent precedents at Cambridge for electing a professor of barely twenty-eight years of age. Glazebrook, who had done much under Rayleigh to build up the school of experimental physics, had hoped for the post, and I think that Rayleigh would at that time have considered him to be the safer appointment. 'My doubt', he said, 'was whether Thomson should be professor of *experimental* physics. He had done very little experimenting at that time, though enough to show that he could do it. But he has shown since that it was right to appoint him.' Glazebrook and W. N. Shaw, who had been Lord Rayleigh's demonstrators, continued under Thomson, who was grateful for their help, realising, perhaps, that the promotion of a man younger and of less experience in experimental work than themselves, could not have been wholly agreeable to them.

Halifax. December 25th, 1884.

Dear Thomson,

Forgive me if I have been wrong in not writing before to wish you happiness and success as Professor. The news of your election was too great a surprise to me to permit me to do so. I had looked on you as a mathematician, not an experimental physicist, and could not at first bring myself to regard you in that light.

However, I think now on Christmas Day I can wish you prosperity. Shaw told you, I believe, about the assistant....

I should like to know your wishes as to next term. I am willing if you desire it to go [on] as I have been doing for a time and to settle during the term what my position and work is to be in the future. I think such an arrangement would be the best for the students at present and I myself hardly feel in a position to make any permanent agreement. Time may bring a different aspect on the face of things.

Yours very truly, R. T. GLAZEBROOK.

The Owens College, Manchester
(Not dated.)

My dear Thomson,

I send you my best and sincere congratulations on your appointment as Cavendish Professor.

I have beaten you once, you have beaten me now in return, and so we are quits.

I have no doubt the laboratory will flourish under your superintendence.

Yours very truly, ARTHUR SCHUSTER.

Heaton Road, Withington, Nr. Manchester.
Dec. 26th, 1884.

Dear Schuster,

Many thanks for your very kind congratulations. I can hardly realise my position yet as I never regarded my candidature as a serious one, and should have been very pleased if either you or Reynolds had been elected. I am glad to say that Shaw is staying on as demonstrator. I have not heard from Glazebrook yet. I hope you wont mind if I bother you with questions about laboratory work sometimes, as I feel I have an immense deal to learn. Wishing you all the compliments of the season, I remain, Yours ever,

J.J. THOMSON

The following was from Osborne Reynolds, his former teacher of engineering at Owens College, who as J.J. implies in the previous letter was a rival candidate:

Fallowfield. 26th Dec. 1884.

My dear Thomson,

I do not like to let the occasion pass without offering you my congratulations, which are none the less sincere that we could not both hold the chair. Your election is in itself a matter of great pleasure and pride to me as it must be to all those connected with the commencement of your brilliant career, and I have no doubt but every hope that you will amply justify the wisdom of the election.

Believe me yours sincerely, OSBORNE REYNOLDS.

10 Osbourne Rd, Clifton. 23rd Dec. 1884.

Dear Prof. Thomson,

I congratulate you and Cambridge on your appointment to the Cavendish Professorship. I was very much afraid they might appoint one of the senior candidates such as J. C. Adams,* or Garnett. Although I would of course have liked to be myself appointed, I think the electors have done very much better in appointing you. I feel that it is betraying my vanity to speak as if I had a chance against you, but by referring to my wish to be appointed I don't intend to infer that I had.

I was afraid they might have thought you too junior but I must now express my hopefulness for Cambridge when it does not consider the most important of all qualifications, namely the energy of youth, as a disqualification. If in no other respect, and as a matter of fact, in very many other respects, you have the advantage of me in that, and I hope that those in authority in Cambridge will always consider every year above thirty as a serious disadvantage to its possessor. In a few years hence with your new statutes some such principle will be of great consequence to the welfare of Cambridge.

I have been over here for the last few days stopping with Prof. Ramsay. I was away with him in the Yellowstone Park this summer and we became great chums. If you have a chance you should try and become acquainted with him and Mrs Ramsay who is perfectly charming.

I hope you will succeed in getting your experimental test of Maxwell's theory tried. The great difficulty is something to *feel* these rapidly alternating currents with. Would Langley's bolometer do? I was working at a receiver whose period of oscillation should be the same as that of the current and which would consequently 'resound' to the vibration and integrate the energy of a large number of vibrations.

Again congratulating you and Cambridge on having escaped all older fogies than yourself including in that myself, believe me,

Yours sincerely, GEO. FRA. FITZGERALD.

The following letter shows that Thomson was already interesting himself in psychical research. (He has given a chapter on this subject in his *Recollections*; the conclusions are for the most part negative, so far as his personal experiences went, though he does not sum up definitely against the phenomena.)

Hillside, Chesterton Road, Cambridge.
Dec. 24th, 1884.

My dear Thomson,

When I met you on Monday it did not occur to me that my anxiety with regard to your career—in connexion with our unpopular investigations—could be so soon shown to be superfluous.

* (Sic) J. C. Adams, the discoverer of Neptune, was already Lowndean Professor of Astronomy. His brother, W. G. Adams, was probably meant.

I heartily wish you success in the important work that has been entrusted to you by so distinguished a body of judges.

Yours very truly, H. SIDGWICK.

To (SIR) ARTHUR SCHUSTER: *Trinity College, Cambridge.*
March 1st, 1885.

We are doing very well at the Laboratory, we have got 97 students doing practical work this term. What an excellent book Stewart's [on Practical Physics] is. He must have taken an awful lot of trouble over it.

Thomson as professor had of course lecturing and administrative duties to occupy him. He had little knowledge of mechanical processes and technique and was at no time ready with his hands. Subsequent to his appointment his experimental work was always carried on with an assistant, and he had in early days the help of two men, both of whom had exactly the qualities in which he felt himself to be lacking. These were Richard Threlfall* and H. F. Newall,† both of whom had engineering ability, and manual dexterity, and for both of whom he had the highest regard and appreciation. His regard was fully reciprocated. After Threlfall's appointment to Sydney he induced Newall, who had been the first undergraduate who ever worked in the Cavendish Laboratory, to return to Cambridge in 1885. He was a son of R.S. Newall, F.R.S., a wire rope and submarine cable manufacturer in the North, and the initiator and donor of the 25-inch refracting telescope of the University of Cambridge. H. F. Newall came at first as a salaried private assistant to the Professor, on the understanding that he was to have time free for investigations of his own. Afterwards he became successively assistant demonstrator and demonstrator, before he left to turn his attention to astronomy. He writes:

J.J. was very awkward with his fingers, and I found it very necessary not to encourage him to handle the instruments! But he was very helpful in talking over the ways in which he thought things ought to go.

* Afterward Professor of Physics at the University of Sydney. Created G.B.E.
† Now Prof. H. F. Newall, F.R.S., Emeritus Professor of Astrophysics in the University of Cambridge.

When Thomson began to turn his chief attention to the discharge through vacuum tubes, it was indispensable to have the services of a glass-blower to make and repair the apparatus used. Lamp glass-blowing is an entirely different craft from pot glass-blowing, which is used for making table ware, or glass globes or tubing. The lamp glass-blower starts with commercially made glass tubes or bulbs, and softens them locally with the blowpipe so as to join or combine them in any manner that may be necessary. This art is by no means a new one—the oldest examples of it that I have seen are the thermometers in the collection of the Accademia del Cimento preserved in Florence, date about 1660. At the time we are writing of, nearly all scientific lamp glass ware was imported from Germany, and although it was possible to get work done to order in London, workmen who could do it were scarce. In any case to wait for such work to be *done out* was an intolerable obstacle, as Thomson and others before and after him have found. English men of science mostly relied upon amateur work done in the laboratory, often with their own hands. Crookes, for instance, worked in this way. In Thomson's early days as professor, the mechanical assistant at the Cavendish was D. S. Sinclair, who was sent to have some lessons, and quickly became very good at it. He left in December 1886, and Thomson's work was very much held up for want of any successor. It is not a natural aptitude with everybody, and efforts to have another assistant trained came to nothing. Thomson heard that a boy in the Chemical Department ('Ebenezer') had some skill, and employed him out of working hours to make the urgently wanted apparatus. Finally, he went over to the Chemical Laboratory, and offered to engage him as his private assistant. Liveing, the Professor of Chemistry, made no difficulty and indeed advised him to accept; and thus began what proved to be a lifelong association. 'Ebenezer' was promoted to be 'Everett' and was established in the room at the east end on the ground floor. This room had originally been called the 'Magnetic room', being designed, I believe, by Maxwell for experiments in which there would be no magnetic disturbances from iron fittings or pipes. It was there that Rayleigh had conducted his determination of the ohm, and there that Thomson's most important researches were carried out. Everett had his blow-pipe there. He sat at it on a high stool which professional glass-blowers (so far at least as I have observed) never do. Though he was chiefly self-taught his work was very effective for what J.J. required. His time was chiefly occupied in setting up the arrangements for Thomson's experiments, exhausting the vacuum apparatus and the like. The assemblages of tables and stands which he made were of a random character and appearance. When I made some criticism to that effect, and suggested that a different arrangement would have been better, he replied, somewhat tartly, that the Professor would not wait for that and expected things to be done quickly.

Thomson did, in fact, as he himself told me, find delays and obstacles very trying when he was on fire with an idea that he wished to explore. I do not think, however, that he ever allowed impatience to get the better of him, or visited the perversities of his apparatus on his assistants who, as he doubtless realised, were doing their best. Once when he came to me on his daily round and I unfolded a tale of woe, he said: 'I have been struggling with broken glass myself for a week past. I believe all the glass in the place is bewitched.' However, it is possible to find a more rational-istic explanation. In those days it was difficult to get glass of uniform composition. Comparatively few laboratories did any difficult glass-blowing and the suppliers of tubing did not turn over their stocks at all rapidly, nor did they always replenish them from the same source. The result was that much of the glass was deteriorated by age and one piece would not fuse satisfactorily to another of different composition.

Everett did not work only for the Professor; he made glass apparatus for other workers in the laboratory, and gave lessons in this indispensable art. His scientific knowledge was limited, and he did not always understand what was the object the Professor was pursuing. For this reason, he was hardly qualified to take responsibility for a series of experimental readings, or for making numerical reductions. Nevertheless his help was invalu-able. He continued with Thomson almost to the close of the latter's career. We shall meet him again in later chapters.

All are agreed that J.J.'s attitude towards the laboratory staff, from the demonstrators down to the laboratory boys, was most kind and friendly. Thus Prof. L. R. Wilberforce,* at that time one of the junior demonstrators, had been brought into accidental contact with a case of smallpox. He writes:

I thought J.J. ought to be told as he might think it inadvisable for me to mix with laboratory students without undergoing some kind of quarantine, and I shall never forget his great kindness. He told me to keep quiet and not to worry, sent me to his own doctor to be vaccinated and observed, and made everything easy for me.

Previous to J.J.'s appointment as Professor all experiments requiring an electric current had to be done with primary batteries. As he himself wrote in 1931:

The most vivid impression I have of my early work in the laboratory is that of Grove's cells; these had platinum foil immersed in nitric acid for one electrode, zinc in dilute sulphuric for the other, and what with the fumes which assailed one's throat and the acid which destroyed one's clothes, the assemblage of a battery of cells was a most disagreeable business. I have not seen a Grove's cell for forty years, and I do not want to see another.

There is a ground-floor room on the left-hand side of the archway entrance which had been designed by Maxwell as the 'battery room', the idea being that the Grove's batteries would be set up and used there. Wires from this room hung untidily all over the laboratory. J.J. now installed a battery of accumulators in this room. A dynamo driven by a 2 horse-power gas engine in the workshop adjoining was arranged to charge them, and a proper and permanent wiring system put in all over the building. Moreover the same engine was arranged to drive the lathes in the workshop. The (comparatively) modern accumulator was invented by Faure in 1881, and allowing a little time for commercial development, the laboratory was fairly up to date in the matter. J.J. wrote: 'I am going to try home made storage cells instead of Grove's for the demonstrations. They use them at the Paris Observatory. They are made just of sheet lead painted with oxide. I have as yet no practical experience of them.' It is not very clear

* Afterwards Professor of Physics in Liverpool University.

whether these constituted the main battery or were merely supplementary.

After Threlfall had left for Australia, J.J. kept up a regular correspondence with him for some years, which has fortunately been preserved. Some extracts from it may be given here.

To R. THRELFALL, *April* 13*th*, 1886,

from Brondem, Colwyn Bay, North Wales:

I cannot tell you how sorry I am that you are going; every day I spend in the laboratory when you are not there makes me feel that on Monday when we elected you [to the Professorship in Sydney] I did the worst day's work for myself that I ever did in my life.

Trinity College. Oct. 24*th*, 1886.

I daresay you have seen in the papers that we have lost our Master. We were all hoping that Rayleigh would be appointed as I believe he would have taken it, but I have just heard that it has been offered to Butler, the late headmaster of Harrow.*

Trinity College. Nov. 14*th*, 1886.

We have been having tremendous discussions with Stuart† about the New Tripos. He wants to run it so that anyone who goes to his shop and learns filing and fitting and a little (very little) mathematics may get the highest honours without knowing any physics at all. We naturally object, and the consequence is we have the most tremendous rows at the Physics and Chemistry Board—What will ultimately come of it I cannot say.

Trinity College. Feb. 3*rd*, 1887.

I am pretty busy this term as Glazebrook is not strong enough to do his demonstrating and I am having to do it for him. . . .

I went to see Gilbert and Sullivan's new piece, Ruddy-gore, last Saturday; it is a strange mixture of good and bad. The first act is, I

---

* Dr Butler was in fact appointed. It was not in Rayleigh's character to decide on what he would answer to any proposal until it was actually made. He said much later that it would have been creating a precedent to appoint anyone not in Orders, and that he did not think the Prime Minister, Lord Salisbury, could well have made such a precedent in favour of a relative of his own. He said, moreover, that he should not have known whether to accept the Mastership or not, and was glad to be spared the decision.
† James Stuart, Professor of Mechanism, 1875–89. Afterwards entered Parliament as a liberal.

think, the best first act that they have done, while the second is by far the worst thing they have ever produced; it is only relieved by a delicious piece of fooling between Jessie Bond and Barrington, dressed respectively as a district visitor and a churchwarden, in these costumes they go through the harlequin-columbine business of the pantomime to slow music.

Our Master is distinguishing himself by the guests he brings to the Hall.* The other night he brought a man who is going to put him in the 'World' as a celebrity at home and who if report speaks truly has been through most of the courts of Europe for immoral practices. The other guest on the same evening was Augustus Harris of Drury Lane. He seemed to find a difficulty in expressing his sentiments in ordinary language and I heard afterwards it was because in ordinary life he was accustomed to the use of language which would make Sedgwick's hair curl.

Trinity College. April 24th, 1887.

I have had a row with Ostwald in the *Phil. Mag.* He attacked my paper on the chemical combination of gases in a book on Chemistry he has just published. He admits now I believe that he was quite wrong, which is satisfactory so far as my theory goes, but does not give one a high opinion of the care he has bestowed on his book.

June 19th, 1887. Trinity College.

Thank you very much indeed for the magnificent things you have sent me from New Zealand. They are the most remarkable things I think I ever saw, and ever since they came, my rooms have been crowded with people to see the 'New God that Thomson has got'. I do not know whether you noticed the head of a sort of devil on the lower part of it. It is the best head of a devil I think I ever saw and I burst out laughing every time I look at it. . . ..

To Mrs H. F. REID:

Trinity College. Nov. 4th, 1886.

I think you would be amused if you were here now to see my lectures—in my elementary one I have got a front row entirely consisting of young women (some of them not so young neither, as somebody says in Jeames' Diary) and they take notes in the most praiseworthy and painstaking fashion, but the most extraordinary thing is that I have got one at my advanced lecture. I am afraid she does not understand

---

* Here we see J.J. in his most paradoxical vein, as anyone who knew Dr Butler will appreciate.

a word and my theory is that she is attending my lectures on the supposition that they are on Divinity and she has not yet found out her mistake.

J.J. seems to imply in this letter that it is not likely that a woman would be able to take advantage of advanced instruction. That opinion was probably fairly general at the time, but only four years later Miss Fawcett was Senior Wrangler, and the position became rather difficult to maintain.

To R. THRELFALL:

Trinity College.   Aug. 7th, 1887.

There is a great agitation going on to admit women to all the University privileges that men have and it seems a most ill-advised thing as it would involve their sitting on boards, etc. and it has divided the friends of the women* nearly as much as Home Rule has divided the Liberals. Sedgwick† in his usual vigorous way, declares that unless they drop the agitation he will turn all the women out of his laboratory, and not allow them to attend his lectures. Is yours a mixed university? . . .

Dec. 11th, 1887.   *Trinity College.*

We are just at the commencement of a great attack which is being made by the supporters of women to secure for them full admission to all the privileges of the university, such as that of voting in the Senate House and serving on Boards, etc. Most of the Residents are most strongly opposed to it, but the non-residents seem largely to take the other view.

I do not think myself that it would do the university very much harm, but it seems to me that it would be bad for the women to be tied hand and foot to our system, for from what I see of them at the laboratory I am sure they require rather a different course from the men: for example they always do very well in the first [part] of the tripos, but make a most awful hash of the second, in fact I think in nineteen cases out of twenty they had much better not attempt [it].

I have been trying lately to measure resistances by determining the logarithmic decrement of a disc oscillating between the poles of a powerful electromagnet. I find I can make the method work all right and I am going to determine the temperature co-efficient of a lot of mercury tin amalgams with it.

* Of whom J.J. was one.
† Adam Sedgwick, the younger, Reader in Animal Morphology and afterwards Professor of Zoology.

Oct. 3rd, 1888.

After a full trial of the methods for testing Maxwell's theory I have come to the conclusion that the most hopeful way is to test whether the electrostatic potential is propagated with a finite velocity or not. If it is then if you have concentric spheres the potentials of which change their signs millions of times in a second, the electric force will be finite outside the outer and inside the inner spheres instead of vanishing as it does if Maxwell's theory is true. By using electrical oscillations you can get the required rapidity of reversal (see Hertz, *Wied. Ann.* 1887) and the only thing is to get some way of detecting an electrical force which is continually changing its direction. Perhaps this might be done by the glow produced in a highly exhausted space such as an incandescent lamp, or since the force is not uniform by the mechanical force it exerts on a small unelectrified conductor. If you care to undertake it, it is a thing eminently worth doing though very difficult. Most of the difficulty would vanish if you could get a delicate way of detecting an alternating electromotive force. I have suggested the one I think most hopeful but perhaps you may hit upon something better.

You should read Hertz's papers; they are very interesting. He has measured the velocity of propagation of electrodynamic action, but this is not the real point which is whether all circuits are closed or not; if they are then the electrostatic potential is propagated with an infinite velocity, but if they are not then the velocity is finite and could be tested in the way I suggest.

6 Scroope Terrace. April 18*th*, 1891.

I am at present very much interested in the theory given in the paper in the *Phil. Mag.*, which is a kind of molecular theory of the electromagnetic field, the tubes of electrostatic induction corresponding to the molecules and all the effects of the field being explained by the motion of these tubes just as the properties of a gas are explained by the motion of the molecules. I find I get a much clearer idea of electrical effects by the aid of this theory than I can from any other. I cannot tell however whether other people will find the same.

Oct. 13*th*, 1892.

I think myself that the *Philosophical Magazine* is a better means of publication than even the Royal Society as the circulation is larger and the delay very much less. I only send papers to the R.S. occasionally as it is usually so long before they are in print that one almost forgets what they are about.

R. THRELFALL *to* J. J. THOMSON, *from Hoffman House, Broadway, New York. April* 13*th*, 1889:

I have been to Edison's place twice. Edison is a capital fellow himself and looks rather like pictures of Napoleon 1st. All one hears about his working three or four days and nights at a stretch is quite true: he has about 100 assistants, and manages, I hear, to keep them working all night too very often. His laboratory is a splendid building like a factory and very like one, and his store room is filled with samples of everything that has ever been manufactured in the way of material. He tells me that in (/76 I think it was) he did a lot of things that Hertz has been doing (he told me some details) and thought he had discovered what he calls 'etheric force' (he is quite ignorant). These experiments he published in the 'New York Tribune' of all places in the world (July 1876) and was prevented from following it up by Silvanus Thompson telling him that it was merely 'common' induction. He seems to have used a detector very like Hertz' with the spark points under a microscope. I will try to get a file of the Tribune and follow the matter up. Edison to this day has not the least notion how things work. In fact I don't know what most to wonder at—his modesty or his ignorance, but anyway he is a charming person. . . .

. . . [Brown and Sharpe's] best thing is their milling machine and grinding machinery. If ever you want one get this. I am going to get one, the thing will do everything but talk. By the way, Edison's new phonograph is ridiculously successful. I never knew what a disagreeable voice I had till I tried one of these machines: there is a big works where they make them; the workmanship is stated to be perfect and *is*. . . .

J. J. THOMSON *to* MRS H. F. REID:

Trinity College. Sept. 26*th*, 1886.

I was fortunate enough to be near a place [in Wales] where there was a wonderful collection of curiosities. . . of a kind that are very difficult to come across. The house was in a country place, and a local auctioneer sold—he was the best seller I ever saw in my life. The way he egged on the bidders against each other was most amusing to watch. He would say 'Now Sir James,* say the twenty pounds, that will frighten him' and then when Sir James had said it he would turn round to another bidder, and say 'Now Captain Jebb, I never saw you beaten before. Say guineas, you will never feel the difference.' Captain Jebb by the by had been weak enough to bring his wife to the sale, and as

* The names have been altered. R.

he was a mild young man, and she a sulky ill-tempered creature, he had to get everything she wanted whatever the price. . . .

The pictures [I bought] are very quaint things: they are on copper and represent the battle of Lepanto, a battle between the Christians and the Turks about 1570; the sympathies of the artist were evidently with the Christians for he has represented their dead disappearing as angels, while the Turkish dead make their exit as devils. . . .

There is a story about an American small boy going about which is rather amusing, and since it illustrates the difficulty of scoring off that specimen of humanity it may, I hope, be called instructive. It is as follows:

Mamma (to small boy who has been lying). Tommy, did you ever hear of Ananias and Sapphira?

Tommy. Hear of them, Ma? I knew them both.

Mamma (severely). Oh Tommy, do you know where they went for telling stories?

Tommy. Yes Ma, I saw 'em go.

The story at the end of this letter has seemed relevant to the subject of this book because it is very typical of the kind of humour which pleased J.J.

From PROF. G. F. FITZGERALD *of Trinity College, Dublin*:

Thorncliffe, Monkstown, Co. Cork.
23rd August, 1886.

My dear Thomson,

I am looking about for excuses not to go to Birmingham and the last sufficient reason for going is because you wrote to me about bringing on a discussion about the Electromagnetic Theory of Light and I said I would help to confound the Jelly-theorists but at the same time I was very uncertain from the way you wrote whether you certainly intended bringing it on and if you have given up the idea or do not still ask my concurrence I will gladly accept the opportunity of a continued vegetation in the lazy atmosphere of Cork rather than the feverish activity of a Brit. Assn. meeting in Birmingham. If you don't get this in time for me to get your reply by Monday morning the 30th don't bother to answer as my fate will be decided without the assistance of your wisdom.

I have a letter half written to you for months on the vortex atom theory of gases but my idle brain has succumbed to the temptations of its natural feeble laziness and has not decided what becomes of the irrotational energy in adiabatic expansions. However, I am afraid my poor fallow mind is growing nothing but weeds in its middle life, in a country where it is proposed that 'Nobody shall pay anything to anybody for five years'; where there is every prospect that education will be put under the thumb of the hierarchy of Rome to 'fetter the intellect and enslave the soul' and where everybody who can at present do anything educationally thinks that Greek verbs are more important than the laws of God. It's simply sickening and I am worse than all the rest for I am too lazy to do anything in this hotbed of fermenting corruption.

Yours sincerely, GEO. FRAS. FITZGERALD.

The following characteristic letter from Sir William Thomson (not yet Lord Kelvin) apparently covers a communication from Mr Oliver Heaviside on some point of electromagnetic theory.

Netherhall, Largs, Ayrshire. Jan. 15/88.

Dear Thomson,

Will you look at the enclosed and deal with it according to its deserts. If it is to be published in Feb. you should send it as quickly as may be direct to Dr Francis, Red Lion Court, Fleet Street, London, with or without a note or remarks or short separate paper wh. you might write to let appear with it. I think O.H. is right in × but his + is unintelligible to anyone who has not read all O.H.'s papers, and it and everything else would be unintelligible to anyone who had. No brains would be left.

In p. 313 of *London Math. Jour.* June 10/86 last line but one from foot, I think dH/dt wants $\mu^{-1}x$ prefixed, and corresponding modification in l. 8 from foot.

p. 321, l. 5. For $4\pi\mu$ subs. $2\pi\mu$
and 9° ϑ° in l. 7?

O.H. rails at the most innocent of J.C.M.'s abstrusities, and loses the benefit of the thing signified, substituting worse abstrusities than Maxwell's worst; and he keeps Maxwell's VERY WORST 'Curl'.

Yours, W. THOMSON.

There can be no possible objection to deferring the Note till March as it arose altogether out of Editorial on a Feb. paper.

J.J.'s reply to this is not to hand, but like Lord Kelvin he was in general impatient of obscurity, and disinclined to take the trouble to follow authors such as Oliver Heaviside who used unconventional methods in mathematics—it would be easier to do it over again, he said.

It was not very long after his appointment as Cavendish Professor that Thomson dined with Sir George and Lady Paget, and sat next Miss Rose Paget, a daughter of the house. Sir George Paget (1809–92) was Regius Professor of Physic in the University, and had been eighth Wrangler in his time. It so happened that the fellowship to which he had been elected at Caius College as far back as 1832 required the holder to study medicine or physic as it was called, and this was what led to his taking up a medical career, in which he achieved high success.

Miss Rose Paget had to some extent inherited his tastes, and feeling the need of intellectual food more satisfying than French and German she gained a fair acquaintance with elementary mathematics, and took the Higher Local Examination. Although not entered at Newnham she attended some classes there. In 1887 she attended the elementary lectures and demonstrations at the Cavendish Laboratory, and in 1888 J. J. Thomson's own lectures, and the more advanced demonstrations. It was his habit to look in from time to time at the demonstrations, and say a few words to each of those at work. In 1889 Miss Paget did some research work at the Cavendish Laboratory on soap films thrown into stationary vibration by sound, after the fashion of Sedley Taylor's phoneidoscope. J.J. found that she needed some help in these experiments. He was working at that time on his determination of 'v' with G. F. C. Searle. They used to arrange to begin work at say 4 p.m. J.J. would come in, and say 'I must go upstairs for a few minutes', which very easily expanded themselves, by 'Relativity' perhaps, into an hour. One day he came down looking highly delighted, and Miss Paget went out with a flush on her cheek, and did not continue any more experiments. This was towards the end of 1889. After six weeks' engagement they were married on January 2nd, 1890. The wedding was from the bride's home at 2 St Peter's Terrace, and took place at Little St Mary's Church.

After ten days' absence, including three days spent with J.J.T.'s mother, they set up house at 15 Brookside, Cambridge, but it was only available for six months. At the end of that time they moved to 6 Scroope Terrace, which had just been vacated by Dr Westcott on his acceptance of the Bishopric of Durham. The house belonged to Caius College. There they remained for nine years.

Immediately after their marriage the Thomsons' home became a social centre for a wide circle of friends, but more particularly for workers in the Cavendish Laboratory. At first breakfast parties were given for them, but later Mrs Thomson had 'at homes' on Saturday afternoons, and usually also on Sundays, when J.J. was there also. In those days there were comparatively few ladies in Cambridge, for the rule that fellows of colleges vacated their fellowship on marriage had only been abolished a a few years earlier, and the effect of the change was not yet fully felt. Mrs Thomson had therefore not infrequently to sustain singlehanded the burden of entertaining a large circle of male callers. Every time the bell rang she hoped for the help of another lady, but often enough the hope was in vain. Sometimes, however, there was help on the male side. Sir Thomas Wade had been our ambassador in China till 1883, and on his retirement had come to live in Cambridge, where he was after an interval made Professor of Chinese. He would often come, and was willing to hold forth in a quiet way. In 1894, when the Chinese-Japanese war was in progress, his views were naturally listened to with special interest. He had no confidence that a Chinese war-ship would be found to carry ammunition of the right calibre for her guns. Another helpful visitor was Dr George Kingsley, brother of Charles Kingsley, widely travelled and a brilliant conversationalist.

During the period of his married life, it was always the custom of J.J. Thomson, as of most of the married fellows, to dine in Hall at Trinity on Sunday nights. After dinner he did not sit over wine in the large combination room, but preferred the less formal atmosphere of the small one, where some of the fellows stood or sat about and talked or looked casually at the current periodicals. If he had scientific friends staying for the week-end he brought them with him.

We have already mentioned J.J.'s early determination of 'v', the ratio of the electromagnetic to the electrostatic unit. He was not satisfied with it, and towards the end of 1888 he asked

G. F. C. Searle of Peterhouse, who had lately become an assistant demonstrator, to help him with a repetition of it. The method was the same in principle, but the vibrating contact key was discarded and a rotating commutator substituted. This gave much more satisfactory results. Searle maintained the speed of this by a stroboscopic method, and Thomson observed the galvanometer and manipulated the resistance boxes—and got a good balance much more satisfactorily than Searle had expected.

When it came to reducing the results, Searle used to go to J.J.'s house—he was by that time married. They sat down, each with a table of logarithms by a different author, and they worked out all the multiplications and divisions independently. The value which they obtained was $2 \cdot 997 \times 10^{10}$. The mean of results by later workers is $2 \cdot 998 \times 10^{10}$ and the best value of the velocity of light is $2 \cdot 9980$. It appears probable that Thomson and Searle's value was within $\frac{1}{1000}$th part of the truth, and it may have been considerably nearer. No further determinations have been made in this country, with the exception of one by Lodge and Glazebrook, by means of electrical oscillations.

In 1893 Thomson published his *Recent Researches in Electricity and Magnetism*, which was intended as a sequel to Maxwell's great treatise on *Electricity and Magnetism*. Thomson had edited the second edition of Maxwell's book and he felt that it would have disfigured the text to have overloaded it with long footnotes. He therefore decided on this supplementary volume. It contains chapters on Faraday tubes of force, on the Discharge of electricity through gases, on Conjugate functions, on Electric Waves, on Distribution of rapidly alternating currents, and on Electromotive intensity in moving bodies. The book is for the most part a severe one, containing formidable calculations dealing with such questions as the propagation of electric waves along a cable with coaxial sheath as a return. The book, like Maxwell's treatise, is in part a summary of the works of others, and in part original. There is no attempt to bring the subject within the range of students whose mathematical training is inadequate; on the other hand it is not the author's object to display his own proficiency in this direction. The comment of a distinguished mathematician is that J.J. often prefers a simple rudimentary method, even though it be longer, to a method which would use somewhat more advanced processes; in such investigations there is no obvious indication that he is an expert mathematician. Yet it is clear that when the need arose, such as the use in electrostatics of the Schwartz-Christoffel formula for conformed representation of a closed rectilinear polygon upon a half-plane, Thomson would rise to the occasion by some process of his own sufficient for his purpose. To use a phrase which has been applied to his work in other fields 'he got there'.

The chapter of his *Recent Researches* which has probably been most often read is that on the Discharge of electricity through gases—the first detailed account of this subject in our language. This chapter, in contrast to the rest of the book, is mainly experimental, and not difficult to read. Thomson's interest in this subject seems to have been originally excited by the beautiful experiments of Sir William Crookes. I once made some remark to the effect that I liked the style of Crookes' papers. J.J. partially agreed, but said he thought they wanted editing by someone who would have cut out what he (Crookes) probably considered the finest passages!

Thomson began to experiment on electric discharge about the time that he was appointed Cavendish Professor, probably with the instinctive feeling that here was the field that really offered a hope of penetrating the secret of the nature of electricity and its relation to matter. The first step was to make himself fully acquainted with what had been done by others. This he did thoroughly, and the chapter in his *Recent Researches* was the fruit of his study. It contains an admirable summary of the facts to date with many acute remarks. But upon the whole, the present-day reader who consults it will see that the author had not at that date the essential clues which could guide him through the maze of phenomena; it is in the main a descriptive account of the facts, without an adequate clue to their interpretation. However, even an account written at the present day would to a certain extent be open to the same criticism. There were not then many clear lights, and in particular, few openings for really telling quantitative experiments. Most of what had been written by Crookes, Hittorf,

Goldstein, De la Rue, Spottiswoode, and others was purely qualitative and descriptive. Such suggestions as they had been able
to make towards interpretation were vague and unsatisfying.

Much effort was spent by Thomson at this time in attempting
to determine the speed with which the luminosity spread along
a long vacuum tube, many metres in length. I saw the arrangement for this experiment on a visit to Cambridge about 1892,
when I was taken round the laboratory by Glazebrook, but J.J.
was not there at the moment, and I did not make his acquaintance
till later. In the then state of the subject this seemed quite a promising line of investigation, but in the light of later progress the
results were not found specially significant, and in fact no mention
is made of them in Thomson's later books.

The same may be said about a long series of experiments on the
electrolysis of steam by the electric spark. A good deal of mixed
gas was liberated along the length of the spark, and when this had
been got rid of, there was an excess of oxygen at one electrode
and of hydrogen at the other. J.J. was at one time inclined to think
that these gases were liberated in the same quantity as in a volta-
meter included in the circuit, but it was found that the excess of
hydrogen was not always at the cathode, but appeared under
certain conditions at the anode. He probably saw on reflection
that the electrolytic conception could not be sustained. At all
events he wrote no more about it in his later books.

Of much more permanent value were his experiments on electrodeless discharges produced by electromagnetic induction. The
phenomena of electric discharge at low pressure are very complicated, but many of the complications are at the electrodes, and
Thomson's idea was to get rid of these complications by producing
the discharge in an endless ring. This he succeeded in doing by
making the path of the discharge in a bulb of rarified air the
secondary circuit of a step-down alternate current transformer.
The primary consisted of a few turns of wire wound round the
bulb, and a leyden jar was discharged through this primary, thus
setting up electrical oscillations. The discharges obtained in this
way have been useful in spectroscopy, not only for their brightness, but also for the facility with which higher degrees of excita-

tion are obtained in the successive zones, as we proceed outwards from the middle of the bulb. Thomson pointed out the advantages of this kind of discharge for studying the phenomena of afterglow, of which very little was then known. He showed this at the British Association Meeting at Oxford in 1894. It was then that I first saw him, and was introduced to him by my father as likely soon to be his pupil at Cambridge. My father gave me to understand how highly he thought of him.

To (SIR) ARTHUR SCHUSTER:

Cavendish Laboratory, Cambridge.
Jan. 20th, 1895.

Dear Schuster,

I am exceedingly sorry to find that my papers have given you the impression that I wished to slight your work. I can assure you that nothing is or has been further from my intention. I will confess that I am not as well acquainted with your second Bakerian Lecture as I ought to be. It was published when I was very busy writing my *Recent Researches.* I had practically finished the part relating to the discharges through gases; when I read your paper it was with reference to the insertions it would require me to make and I no doubt passed over the part of your paper relating to parts of the subject which I had not introduced into the Chapter on Discharge through Gases too hurriedly. I cannot, however, plead having forgotten Stanton's experiments on the discharge from red hot copper for I had them in my mind when I was writing the paper in the Dec. number of the *Phil. Mag.* and I fully meant to have referred to them, it was a mere accident that I did not. I am writing by this post to the Editor of the *Phil. Mag.* in which I refer to his experiments and point out that you more than five years ago stated that the negatively electrified atoms moved faster than the positive. If I had known of your results earlier, it would have saved me a great deal of time and worry, for when some of my experiments seemed directly to suggest it I had the greatest repugnance to the idea and fought against it much longer than I should have done if I had known it had commended itself to you. With reference to the third point I am not quite sure as to which paper you refer. The only reference I remember to have made on the subject of the deflection of the cathode rays by a magnet was at the end of a paper on the velocity of these rays when I made a rough calculation as to whether the deflection by a magnet was compatible with the velocity I had found. I certainly did not know until I received your letter that you had made a series of measurements of the deflection of the rays by a measured magnetic force, and I should not have thought of looking in your first Bakerian Lecture as I thought I did remember that paper, and that it was about the discharge through mercury vapour and the residual effects observed near the poles when the current was reversed.

I can only express my regret that my ignorance should have caused you any annoyance.

Believe me,

Yours very truly, J.J. THOMSON.

It was in 1893 that the 'Cavendish Physical Society' first met. This was not a society in the ordinary sense of the term, for there were no list of members, no subscriptions, and no publications. It was rather what is called on the continent a colloquium. In Felix Klein's *History of Mathematics in the Nineteenth Century*, p. 113, it is stated that Jacobi founded the first Mathematical Seminar in Koenigsberg in 1834. The idea seemed so strange to the people there that Bessel, the astronomer, refused to take part. Since that time the institution has become fairly widespread on the continent. It is believed that J.J. was the first to arrange anything of the kind in this country. It was held fortnightly on Tuesdays during term time, in the lecture room of the Cavendish Laboratory, with the Professor in the chair; and anyone interested could come in. From the autumn of 1895 onwards there was tea before the meeting, and this was presided over by Mrs Thomson, whose idea it was to give it. She occasionally brought in other ladies to help her. J.J. was anxious that the meeting should not degenerate into a social function, and his idea of guarding against this was to insist on the plainest possible cups and saucers! The institution of tea was, however, a good one, because many of the workers in the laboratory had had little or no lunch, and required some sustenance if they were to take effective part. The actual workers in the Cavendish were notified by a paper on the laboratory door, but circulars were sent to some other likely attendants. Sir George Stokes, for example, not infrequently put in an appearance.

Some member of the teaching staff, often the Professor himself or a research student of reasonable maturity, or sometimes a visitor from another department, such as Prof. Liveing, or Prof.

Ewing, would either give a talk about something of his own, or expound some interesting recent paper, English or foreign, showing experiments when possible. After the paper there was a discussion. J.J. was very anxious to keep it going, and as he was much readier of speech, and usually knew much more than anyone else present, this naturally resulted in his doing most of the talking, alternating with anyone else who had enough self-confidence to debate publicly with him. I do not mean to suggest for a moment that J.J. wished to dominate. What happened was that the discussion began to flag because other people were shy or inarticulate, or had not collected their ideas, and that he had a strong and justified impulse to prevent it dropping. Later, when there were several American professors working in the laboratory, the balance became more even, and the discussion more general. The attendance varied a good deal. When a new discovery of popular interest, such as argon or the X-rays, was to be expounded, the room would be crowded to overflowing. Apart altogether from the interest aroused there can be no doubt that these discussions were of value in giving young scientific men practice in exposition and in debating.

On the whole Thomson's earlier studies on electric discharges did not for a good many years reach the level of interest and importance of his later ones. The great turning-point was the discovery of the X-rays by Röntgen in 1896. We shall see presently in detail how this led to a great outburst of activity in the Cavendish Laboratory. Although this, and the Research Student Movement, at about the same time, marked the beginning of a new epoch, it is not possible to make a clear-cut separation between the earlier epoch and the later. Some topics necessarily cut across the boundary.

The earlier years of his professorship were during a period when, with the exception of Hertz's discovery of electric waves, sensational advances in Physics were somewhat lacking. The work done by Thomson in those years, though not of outstanding importance in comparison with what he was able to do later, nevertheless came well up to the general level of scientific interest which prevailed elsewhere.

As a lecturer to elementary students, Thomson was hardly to be surpassed. He had that rare quality of not going too fast, or trying to cover too much ground: thus he avoided being over the heads of the slower students; at the same time he was never so slow as to bore the quicker ones. Though the main outlines of the subject were fully emphasised, he often introduced speculative ideas which reflected his own recent thoughts or historical matter from his own recent reading. He often showed one or two experiments, but these, though introduced in an appropriate place, were not allowed to occupy the forefront of the picture. The lectures were not an exhibition of showmanship. They were intended to teach serious students and were quite exceptionally stimulating to any intelligent pupil.

His elementary lectures for the October term were on the subjects conventionally classed under the heading of 'Properties of Matter' and for the two following terms on 'Electricity and Magnetism'. I believe he inherited the tradition of giving moderately elementary lectures on these subjects from his predecessors Maxwell and Rayleigh, but of this I am not sure. Speaking for myself I think I learnt more from these lectures than from any others I ever attended. Though they were necessarily in great part repeated from year to year, yet they were essentially extempore and the notes, if any were used, were of a very brief kind. J.J. never got seriously muddled in a calculation on the blackboard, and might be trusted to get it out right. His example in this respect was not one to be imitated by weaker brethren, who are generally well advised to have their calculations written out neatly and correctly to refer to as they go along. J.J. always illustrated his general result by a numerical example, realising that if this is not done, the general result will to many be a mere piece of technical jargon, which the class may be able to reproduce, but which they will not understand. He had the invaluable habit of saying everything over again, in a different form of words. The listeners who find this superfluous are few indeed. I have seen the criticism that J.J.'s lectures were too much like a barrister's brief. Possibly he was open to this criticism when he was discussing debatable matters on the borderland of the unknown; but it has not much application when the object was rather to bring home to the elementary student the established conclusions of science. In such cases definite guidance is generally best. Otherwise the learner may be left halting between two points of view, and perhaps clearly understanding neither. J.J. often used to make excursions into the history of science. He sometimes showed experiments which were outside the ordinary. For example, he usually demonstrated the gravitational attraction between small bodies, using C. V. Boys' apparatus for this purpose. Again, in 1888, when the experiments of Hertz on electric waves were a novelty, he demonstrated these by the original very difficult method, which of course could only be used by one observer at a time, and the interest and enthusiasm of the undergraduates was very great. But perhaps his treatment of more commonplace topics was the most valuable. His lectures on electrostatics ran to the refrain of Gauss' theorem. We were reminded several times in each lecture that 'the total normal induction over a closed surface is 4π times the charge inside the surface', and the whole subject was developed from this principle, applied to simple cases, economising our mathematics, as he said. Similarly in the lectures on electromagnetism the text of the sermon, often repeated, was that 'the work done in taking unit pole round a closed circuit is 4π times the current flowing through the circuit'. This method of repeating a fundamental principle almost *ad nauseam*, and showing its fertility in all kinds of directions, was well suited to the class of students he was dealing with. His textbook, *Elements of the Mathematical Theory of Electricity and Magnetism*, contains the substance of these lectures in a somewhat expanded form, and has run into five editions. It has probably moulded the ideas of many generations of students in these matters, outside of those who were privileged to hear the living voice. He spoke clearly, and in loud and resonant tones. His sentences were well finished, though he was apt to fill up gaps with —er— while pausing to form the next sentence or to find a phrase. I think that in later years he largely cured himself of this defect.

J.J. also collaborated with his friend J.H. Poynting in producing a *Textbook of Physics* which was intended to be complete in six volumes. J.J. wrote most of the volume on Properties of Matter, and probably made some contribution to the Heat and Electricity, but the greater part of these and the whole of the Sound was due to Poynting. The volumes do not use advanced mathematical methods, but subject to this limitation they cover the most important aspects of the subject as then known. The intended second volume on electricity, and that on light, were never published.

An admirable word portrait of J.J. in lecture was painted in a series called 'Letters to Lecturers' appearing in the *Cambridge Review*. The particular 'Letter' appeared on February 20th, 1890. It was anonymous, but was generally attributed to W.G. Clay, of Trinity, bracketed ninth Wrangler 1887, who met an early death by an accident in the Italian Alps. It ran as follows:

Letters to Lecturers

XX. To PROFESSOR J.J. THOMSON

Dear Professor Thomson,

When the scientific history of the present century comes to be written, many ingenious theories will no doubt be put forward, to account for the frequent association during it of the name of Thomson with the possession of the highest mathematical talents. I do not know whether you can throw any light on the matter; at any rate you seem to have easily acquiesced in your destiny, and reconciled yourself to following in the footsteps of your illustrious namesake: indeed I can never sufficiently admire the delicacy of the compliment you paid him, by declining the garish glories of the Senior Wranglership, and contenting yourself, as he did, with the second place.

To some men the critical tide in their affairs comes late in life, but not so with you. Do you ever look back to the occasion when you first tasted (was it surreptitiously, I wonder?) the pleasures or pains of tobacco, and reflect that the whole course of your life was then determined? For had a 'Counterblast to Tobacco' in some form or other prevailed with you then, you might never have sat watching the vortex rings of grey smoke, vibrating as you projected them forth, you might never have written the essay which won you your fellowship at Trinity, and which, afterwards expanded and glorified, gained you the Adams prize and your Professorship, and all that that has brought you. But I am forgetting that it is to you as a Lecturer that I have to write. Let me transport myself to your lecture room at the Cavendish

Laboratory, and take my seat among the class waiting for your arrival, and amusing themselves during the interval—for you know, dear Professor, they have occasionally to wait some time—by speculating on the nature of the weird implements on your lecture table. Presently the quick resolute step is heard, the Professor enters, a hasty and often puzzled glance is cast on the apparatus, the bundle of scraps of paper and old envelopes is deposited on the table, and the lecture begins.

Of your lectures themselves I need only say that they prove you to be worthy of your predecessors, Clerk-Maxwell and Rayleigh, and I can give you no higher praise. It is very apparent that you are always physicist first and mathematician second. For when in the course of some investigation a new function turns up, which would keep some of your colleagues at Trinity contented and happy for months, you merely 'with a grave scornfulness' select such of its properties as you require, and march straight on to the goal you have in view, and this accounts for what sometimes befalls you in lecture room. For though knowing well what is the result you wish to obtain you have occasionally mislaid the envelope-back containing the details of the investigation, and are compelled to plunge at short notice into a sea of symbols. Yet when, since even Professors make slips sometimes, it becomes evident that the desired result is not coming, and you find it necessary to apply an empirical correction to the work on the blackboard the cool confidence with which you say 'Let's put in a plus!' and the smile of cheery conviction with which you turn to your audience, puts to shame the incredulity of the most sceptical among them. And he needs to be well assured of his own ground who would attempt to catch you tripping. Do you remember how in the days when you used to lecture on Statics at Trinity, you were once solving a problem concerning a string, which according to you was sustaining an end pressure instead of a tension, a thing no well-regulated string would endure for an instant without collapsing? Some of your class perceiving this thought their chance had come and began to scoff only to find themselves silenced by your ready rejoinder: 'Well then, let's call it a rod.'

Let me recall to you finally your generous discernment of the merits of others, for not a few have owed fellowships and even professorships to your powerful advocacy, and let me say that the words you used to express your pleasure at repeating in Cambridge the experiments which have confirmed so thoroughly the theory of the great Cambridge Electrician, revealed a side of your character perhaps hardly apparent on a superficial acquaintance.

And what parting wishes, dear Professor, shall I express for your future? You have applied Dynamics to Physics and Chemistry; why not extend the application still further? This is the task I propose to you—to find the Lagrangian Function of the University. Think of the advantage of being able to obtain by a simple mathematical process the report of Syndicates on any matter, for such things, I imagine, would correspond to Equations of Motion. Think of the saving of time, labour and expense. Only do this, and great indeed shall be your fame.

I have only one footnote to add to this account. When the Professor says 'Let's put in a plus' it is pronounced ploos—a relic of his Manchester upbringing.

The introduction of practical physics for medical students put a strain on the available accommodation which it was unable to bear, and the most desperate makeshifts were necessary. At first the practical classes for them were carried on in the lecture room, with the simplest appliances. Later certain corrugated iron sheds were annexed, which had been used as dissecting rooms for the department of Human Anatomy. The number of students rose as high as two hundred and twenty in some terms. A small laboratory boy who was employed there was in terror of the ghosts of the dissected subjects, and was generally found in tears if he had been left alone in the room. The Professor took him in hand and reasoned with him successfully. He seems at this early date to have given promise of the tact which, as we shall see in a later chapter, enabled him as Master of Trinity to deal with tearful mothers.

The mention of laboratory boys calls to mind an incident which (according to J.J.) happened about this time, and which appealed very much to his sense of humour. A lady student from Newnham or Girton fainted one day in the laboratory, and a laboratory boy, anxious to rise to the occasion, thought it right to turn the fire hose on to her!

However, to return to the problem of accommodation. Money was the difficulty, the University having little to give. It did, however, provide a site—the piece of land adjoining the Porter's Lodge running down Free School Lane, and measuring about 100 × 40 ft. Thomson had accumulated a sum of about £2000 from fees, and the University found another £2000. A large room on the ground floor occupied the whole site. Underneath was a cellar for experiments requiring a constant temperature, and above was an additional lecture room, and a private room for the Professor, the old room used by Maxwell and Rayleigh having been absorbed for the elementary demonstrations (a grade higher than the medical classes), which had overflowed from the large class room adjoining. This was all that the money available would run to, but the design was such that it could be added to if and when the expense could be met. The new rooms came into use about April 1896.

The opening of the new buildings in March 1896 was celebrated by a conversazione at the laboratory. The entrance from Free School Lane to the laboratory was covered in with an awning, carpets were put down on the stairs, and everything brightly lighted with electric light, temporary wiring having been put up by the laboratory staff. There were scientific exhibits, not only in physics, but in other subjects as well, Rutherford's magnetic detector for electric waves being perhaps the show piece. Mrs Thomson received the guests holding a bouquet of flowers, the gift of the research students. Various workers in the laboratory, the present writer among them, acted as stewards, wearing badges. As Rutherford describes it in a contemporary letter: 'J.J. himself wandered round looking very happy and grinning at everybody and everything in his own inimitable way.' The party amounted to some seven or eight hundred guests, including most of the leading figures in the University.

The laboratory finance in those early days was very difficult, the principal resource being the fees received from the students and from the colleges for which examinations in practical physics were arranged. Thomson had only been able to save the money for the extensions by a cheese-paring policy almost comparable with that practised by Lord Cromer when he was restoring the finances of Egypt. The smallest expenditure had to be argued with him, and he was fertile in suggesting expedients by which it could be avoided—expedients which were more economical of money than of students' time. Mr W. G. Pye, who was in charge of the workshops, and whose name has since become widely known in connection with the manufacture of radio receivers, pleaded earnestly for a milling machine. Other people saw more clearly the need for a new battery of accumulators, but the Professor considered that the old ones could go on for a time longer. When the resources of verbal diplomacy had been exhausted, it was whispered that he could only be convinced by arranging so that he could not get the current for his own experiments.

Naturally, this financial stringency and the rapidly increasing number of workers in the laboratory created a severe competition for such apparatus as there was. A few who could afford to do so provided things of their own. Naturally the scarcity led to the development of predatory habits, and it was said that when one was assembling the apparatus for a research, it was necessary to carry a drawn sword in his right hand and his apparatus in his left. Someone moved an amendment—someone else's apparatus in his left.

This stringency must have borne hardly on the Professor, as well as on his students, and no doubt he often denied himself what he ought to have had—I can recall discussing with him whether a new form of electroscope for radioactive measurements designed by Curie was worth the £5 that was asked for it! I had seen an example elsewhere and was able to answer that it was. It is doubtful, however, whether science was really very much retarded by this kind of thing under the conditions which then prevailed, when appliances were simpler, and the work of the instrument maker was not nearly so far ahead of the rough and ready constructions of the amateur as it is now.

It is on record that before 1914, when about thirty research students were working in the laboratory, the cost of their researches was about £300 per annum. Even if, as is probable, this figure includes no overhead charges, it is still striking enough and will probably convince the reader that the above picture is not overdrawn.

In the Cavendish Laboratory, it was J. J. Thomson's practice to go the round of the research workers every morning to ask how they were getting on and to suggest how their difficulties could be overcome. I think he usually spent about an hour in doing this, from after his lecture, that is from noon till lunch time. It was easy to follow his progress about the laboratory because most of the doors were open, and his voice at that time was loud and resonant—so much so in fact that some modest people were rather shy of having their conversation with him— or at least one side of it—so publicly advertised. He had a wonderful power of putting aside his own preoccupations to enter into what a student was doing, and to encourage him when necessary. I remember, one day, saying how uphill the work was, and he answered, 'Yes, that is why there is so much credit in doing anything'. He was nearly always ready with his comments. If things were going well he was as pleased as the man himself. If the work had got stuck he would sit on a stool alongside, push his glasses up on his forehead, remain silent for a short while, and then shoot out a suggestion for a modification of the apparatus or something new to try.

This action of pushing his glasses up was very characteristic. He was shortsighted, and wore the glasses so as to be able to see objects at a distance. Without them he put his eyes rather close to the object. This is well seen in the photograph facing page 222, which though actually taken in America, gives a vivid idea of his attitude and expression when he was being shown something by a student in the Cavendish.

We introduce at this point a description written by Rutherford to his fiancée, Miss Newton, December 2nd, 1896, which, although already published in his *Life*, cannot be omitted from an account of J.J.:

> In your last letter you ask whether J.J. is an old man. He is just forty and looks quite young, small rather straggling moustache, short, wears his hair (black) rather long, but has a very clever looking face, and a very fine forehead and a radiating smile, or grin as some call it when he is scoring off anyone.

Apart from J.J.'s regular round in the morning, those who worked on the ground floor had frequent opportunities of talking to him at various times of the day. He had no idea of reserving a time during which he was not to be interrupted. There are of course frequent intervals in experimental work when matters cannot be hurried—a suspected air leak in vacuum apparatus has to be given time to declare itself, or a glass apparatus recently made has to be given time to cool, and so on. The rooms on the ground floor all opened into one another, and their occupants wandered to and fro as they felt inclined. J.J. occupied the room at the end, and for a time his assistant, Everett, tried to establish the convention that it was private, indeed I think there was a notice to this effect on the door. But in practice little attention was paid to it, and when Rutherford, McLennan and others were established to work there as well as Everett, the game was up. Rutherford and J.J. talked on and off at all times, discussing the papers in the latest number of Wiedemann's *Annalen*, or the *Philosophical Magazine*. Later, after Rutherford had left, H.A. Wilson became his chief confidant. I myself began work upstairs, but when I had acquired some seniority among the research workers, I petitioned to be removed downstairs, where the heating was less miserably inadequate. The copper hot-water pipes which had been put in by Maxwell were altogether too small. This material was no doubt chosen to avoid magnetic disturbance, but its high cost had probably led to too small a size.

When I had moved down, I too enjoyed constant intercourse with J.J. There were often triangular discussions between him and H.A. Wilson and myself about the nature of the α particle and similar topics. Later N.R. Campbell and F. Horton enjoyed the same privilege. He would often stay talking to Horton after the place was nominally closed, and other students had gone. He would ask if Horton knew how any newcomers to the laboratory (see next chapter) were settling down to life in Cambridge. Sometimes he discussed any piece of combination room gossip. It was noticed that if there was a piece of scandal going about in Cambridge, Thomson always heard about it first!

The picture that remains of the laboratory in those days is of a score of individuals scattered about in various rooms, two or three in a larger room—but each working at his own particular problem, for there was no team work then. Glass work of very varying quality was usually conspicuous at bench level, with the ubiquitous Töpler pump attached, and a maze of wires overhead: at least they should have been overhead, though I remember making a friendly protest to Townsend, who worked in the same room as I did, against his stretching wires in a position which threatened me with decapitation. The research workers had their own habits for commencing work and finishing it at times suited to their own individual temperaments. Some arrived at noon or even lunch time. There were no strict regulations in force (though this may have been tightened up later) and some I believe even got the key from the Porter's Lodge and came in on Sundays.

J.J. on his daily rounds sometimes made very impracticable suggestions, and this was possibly due to his being overworked. Thus, when Horton was requiring to measure with accuracy the diameter of a fibre 0·001 cm. in diameter, J.J. asked whether he had tried fixing a thread to the fibre, and wrapping it round, so as to find what length was required for n turns. There was, however, a useful idea behind this suggestion, which was to measure the circumference rather than the diameter; Horton was able to carry this out by a method of rolling, measuring the distance rolled for a hundred or so revolutions.

When J.J. was posed with difficult questions, he would sometimes say that he would think it over—but this was comparatively seldom. I asked him once whether in Rowland's experiment on the magnetic effect of electrical convection he considered that it was necessary to divide the disc into sectors insulated from one another, and he said 'I do not know how to answer that'. The incident remains in my mind because it was so very rarely that one drew blank. Sometimes we got into argument with him and although he never resented it if we raised objections to or criticisms of what he said to us, or even of what he had committed himself to in print, and readily allowed us to argue, yet we seldom succeeded in having the last word even when we flattered ourselves that we had established our point. However, if we really had, we found it was tacitly accepted the next day. Sometimes beginners in research work, particularly if they were not very clear-headed, were inclined to think that a description of their experiments, even if they led to no particular conclusion, would form a contribution to science. Thomson might on occasion be overheard explaining kindly but firmly to one of his men that e.g. he had not succeeded in getting the numerical value of some proposed datum, and that therefore he had not arrived at his goal. But though unpleasant truths were told when necessary, his daily round was looked forward to, and the pleasure of a few minutes' private talk with him was highly valued. His genuine and kindly interest in what we were doing was an experience above all price.

When distinguished visitors came to the laboratory, and many such visitors came, J.J. always brought them round, introducing the visitor to each research student and getting him to give the visitor his own account of the work in progress. Some of the visitors seemed shy of asking questions, but one striking exception was Kelvin. He was sure to enter into the minutest details. As one student (R.S. Willows) expressed it: 'Kelvin isn't satisfied till he knows what each block of paraffin and each bit of sealing wax is for.'

J.J., like everyone else in an administrative position, sometimes had to say unpleasant things, and he did not shrink from such duties if they came his way. But he was not given to telling 'home truths' unless his duty compelled him to do so, or to interfering more than was absolutely necessary, though he generally knew well enough when there was any friction or want of harmony.

We have already mentioned J.J.'s work on the determination of 'v', the ratio of the electromagnetic to the electrostatic unit. But his greatest strength as an experimentalist was not perhaps brought out in quantitative work of this kind, where the main object was to improve on the technique of previous workers, and in his own later work he wisely turned away from it and devoted himself to exploratory and pioneering work, where the object was rather to discover general relations between the quantities concerned than to follow these out with the closest numerical accuracy. He was at times rather impatient at the way some of his students, instead of pressing on into the unknown, would go back to what they had recently done, and try to improve the accuracy of their measurements. He did not feel, however, that he could tell them this in so many words. Whether, and to what extent, it is advisable to pause in order to secure additional accuracy is a question to which there can be no general answer. Much must depend on the circumstances of the particular case, and it may be added, on the temperament of the individual. To deprecate in general terms the importance of accuracy was he felt impossible.

Tea at the Cavendish Laboratory was a great institution. I speak here of the everyday tea, not the tea presided over by Mrs Thomson before the meetings of the Cavendish Physical Society. Tea, as I remember it, was in the new Professor's room built in 1896 at the same time as the ground floor demonstration room for the medical classes. I have heard Rutherford claim that tea was instituted on his initiative.

The tea hour was in many ways the best time in the laboratory day. The tea itself had no special quality; the biscuits were unattractive in the extreme, and very dull; but the conversation sparkled and scintillated and as a social function tea was an outstanding success. There seemed to be no subject in which J.J. was not interested and well informed; current politics, current fiction, drama, university sport, all these came under review. The conversation was not usually about physics, at least not in its technical aspects, though it often turned on the personalities or idiosyncrasies of scientific men in other countries, who were known personally to some of those present and by reputation to all. J.J. had something to say on nearly any subject that might turn up. He was a good raconteur, but also a good listener, and knew how to draw out even shy members of the company. His laugh of enjoyment when anyone came to the point of their story or anecdote was very infectious, and put the teller thoroughly at his ease, if he had by any chance felt that he was being too forward. In his best form, J.J., while talking, paced the room vigorously in a manner rather suggestive of a caged lion. He might be criticising the conduct of British generals in the Boer War, or discussing the merits of the English players in the Cricket Tests with the Australians. On occasions his blotting pad indicated his choice of an all-England team to do battle with the Australians. One day the Oxford and Cambridge boat race was under discussion, and Bumstead of Yale mentioned that in America railways ran alongside the river on which the Harvard-Yale race was rowed, thus enabling spectators to follow it from start to finish. J.J. burst out, 'Oh, is that how fast your trains go?' 'No,' replied Bumstead, 'that's how fast our crews row.' J.J. laughed heartily at being scored off in this way.

Although these teas were open to everyone in the laboratory of graduate standing, they were chiefly frequented by the research workers and not, so far as I remember, by those who were mainly engaged in teaching. The general conversation might be terminated suddenly by J.J. turning round to an individual and asking, 'How is your experiment getting on?' This was the signal for dismissal.

Although in his middle thirties, J.J. had not yet made the great discoveries by which he will always be remembered, his reputation in the learned world generally was rising. He wrote (April 17th, 1896): 'I have got my hands full just at present. I have to give the Rede lecture this term, preside over Section A [of the British Association] in September, and lecture in America in October.' His Rede lecture (on a famous Cambridge foundation dating from the time of Henry VIII) was entitled 'Röntgen Rays'. It was given to a crowded audience in the lecture room of Anatomy and Physiology, but was not published. The address to Section A of the British Association at Liverpool was also mainly on this subject, and the American lectures given at the University of Princeton, New Jersey, in connection with their 150th Anniversary Celebrations, were on the kindred subject of the discharge of electricity through gases. They were revised for publication, with some additions, in August 1897, and were then published in a small volume. Though now only of historical interest, they form a valuable record of the development of Thomson's views at this date. Mrs Thomson accompanied him. Before the Princeton celebrations they took the opportunity of visiting Professor and Mrs H. F. Reid at Baltimore. Prof. Reid had been one of the early workers under J.J. at the Cavendish Laboratory, and was now Professor of Geophysics at Johns Hopkins University. He and Mrs Reid were among the Thomsons' greatest friends. J.J. gives a full account of his American experiences in his *Recollections*, and there is nothing to add. He says that he got back to Cambridge in time for the October term, but in fact he got back on November 2nd.

It has sometimes been thought that the merits of Prof. Philipp Lenard of Heidelberg had not been sufficiently recognised in Thomson's writings, and that this might help to explain his bitter and almost unbalanced anti-English diatribes during the war of 1914–18. It was certainly not the general impression of Thomson's pupils that he undervalued Lenard's work, and Lenard had demonstrated his experiments at Liverpool by Thomson's invitation. They were fellow-guests of Sir Oliver Lodge for the meeting. In this connection the following letter is of interest, and seems to show that the idea above mentioned is untenable:

Heidelberg. Nov. 20, 1896.

Dear Prof. Thomson,

Let me thank you most heartily for your kind letter and invitation to Cambridge. This award [of the Rumford Medal] by the Royal Society makes me of course very happy, I think it is the greatest event that has happened in my life, but to have so many kind words from you on this occasion makes me still happier. To come to England and to enjoy personal intercourse would be an increased delight to me now....I hope for future occasions to have the pleasure of meeting you. I shall never forget the pleasant meeting at Liverpool and how much of the delight in it was due to you.

Please present my best compliments to Mrs Thomson, and believe me always,

Yours very sincerely, P. LENARD.

The following refers to experiments on electric waves by Prof. J. C. Bose.*

6 Scroope Terrace. Nov. 16*th*, 1896.

Dear Lord Rayleigh,

I have had occasion to read several of Bose's papers and am of opinion, that he is a very suitable person to receive 'Encouragement' if any is going. His experiments are ingenious, and his apparatus very well devised, and it must have required great patience and determination to make the apparatus work in a climate like that of India. The results

* Later Sir Jagadis Chunder Bose, C.S.I., F.R.S.

are very interesting and his paper was received with great applause at Liverpool. His results showed a remarkable agreement with theory, so close in fact that I suspect he has availed himself of the acknowledged imperfections of his apparatus and rejected observations which did not give the right result. This however is only a surmise and I have no doubt that with a well-constructed piece of apparatus of the kind that he has devised very good results might be obtained.

Yours very truly, J.J. THOMSON.

CHAPTER III

GREAT DAYS AT THE CAVENDISH LABORA-TORY. X-RAYS AND GASEOUS IONS

THE antecedent condition for an outburst of intellectual activity, whether aesthetic, literary, or scientific, is an interesting subject for discussion. There have of course been such movements on an enormously larger scale than the one at which we have arrived at this point of our story. The artistic and literary renaissance in sixteenth-century Italy is commonly attributed to the stimulus derived from the rediscovery of the literary and artistic treasures of antiquity. It may be doubted whether this rediscovery was not rather the effect than the cause of intellectual revival. At all events the explanation would not appear to possess the merit of generality; for probably no one would maintain that the age of Pericles, from which the Italian Renaissance is supposed to derive, was itself the result of the rediscovery of a still earlier civilisation. So far as external causes can be assigned at all, it would seem that the conditions prerequisite are intellectual leaders of great power, capable of seizing such opportunities as the state of contemporary civilisation affords, and a public able to appreciate and support their efforts, and to produce men who can benefit by their example and carry on their tradition.

Science has had its periods of renaissance also, though so far as pure science is concerned the movements have been on a smaller scale than those above mentioned. There was a school gathered round Galileo in his later years. There was one in England in the early days of the Royal Society, and, speaking nationally, Newton was its central figure. It does not appear, however, that Newton, in the days of his Cambridge professoriate, exercised a very wide personal influence. His fame spread through his writings. He lectured, it is true, on his optical and astronomical discoveries, and received his pupils in his rooms, so that they might ask for further explanations: but it does not clearly appear that he ever directly inspired younger men of his epoch, such as Cotes, with ideas for researches of their own. Later on, Stokes had given admirable experimental lectures, but it does not seem to have occurred to him to put younger men in the way of experimenting, even when they were particularly eager for it. Maxwell made a beginning in inspiring research work at the Cavendish Laboratory, but his illness and premature death prevented its development. Rayleigh had done more, and had enlisted a number of younger men such as Glazebrook, Shaw, Schuster, J. J. Thomson himself, and a few others to help in his undertaking of revising the absolute system of electrical measurements: but this line of work was too difficult to lend itself very well to the initiation of beginners; and moreover his five years' tenure was rather short to build up a school.

The conditions for doing this are somewhat special. The first is a stimulating leader, one who is not only abounding in energy and ideas, but also one who can without too great an effort throw himself into the difficulties of others. This requires a peculiar kind of versatility not always easily combined with great powers of concentration on any one line of thought. Then again it is necessary to have a productive line of investigation opening up: lastly it is necessary to have the right kind of pupils. These must be men of the not very common kind of ability which makes a scientific investigator: they must not be too young: and they must be provided with the means of subsistence while the work goes on. In earlier times at Cambridge this last difficulty had been serious. The only people available had been the fellows of colleges, and those fortunate enough to be possessed of private means. A change was coming about, however. Some college scholarships were now tenable for a year or two after graduation. There were studentships such as the Clerk-Maxwell and the Coutts-Trotter. Another circumstance was this—the demand for elementary instruction in physics was increasing, notably from the requirements of medical students. This involved the appointment of several demonstrators and junior demonstrators: and the salary attached to these posts, though small, combined with the leisure which they afforded on alternate days, provided for some research workers.

Inner Court of the Cavendish Laboratory, 1909

So much for students who had taken their undergraduate course in the University. Previous to this time Cambridge had not readily opened its doors to post-graduate students from elsewhere. In February 1894 the question was discussed by the Council of the Senate,* and they appointed a syndicate to report, consisting of A. Austen Leigh, F. W. Maitland, Donald Macalister, R. C. Jebb, A. R. Forsyth, J. Armitage Robinson, Alfred Marshall, H. M. Gwatkin, M. Foster, J. J. Thomson, A. W. W. Dale, W. Bateson. There is no written record of their proceedings, and the sole survivor (Forsyth †) does not remember that J.J. took any prominent part, or that the original initiative came from him. The scheme was passed by the Senate on April 25th, 1895, and allowed of degrees for graduates from outside by two years' residence with an approved thesis, or alternatively with a high examination standard. It will be seen that the scheme was not specially directed to science in general, or to physics in particular. In formulating it, the chief constructive part was taken by the Secretary of the Syndicate, J. Armitage Robinson, a theologian, and afterwards Dean of Westminster.

The Commissioners of the Exhibition of 1851 were, and are, charged with administering the funds which accrued from the success of the exhibition, which were invested in land at South Kensington. In 1891 they had founded research scholarships to be awarded to students of selected universities in Great Britain and Ireland, and in Canada, Australia, New Zealand and South Africa. The scholars were elected on the nomination of the ruling bodies of these universities and the scholarships were ordinarily tenable for two years. Until 1896 the holder of such a scholarship could spend his first year at the Institution by which he was nominated, and most of them did so; but in 1896 the rule was altered, and all nominees were required to proceed at once to an institution *other* than that by which they were nominated.

Where then should young physicists go? In earlier years, when only one year abroad was enforced, the German universities had attracted some, but now, by going to Cambridge they could not only get the guidance and inspiration of one of the leading British

* J.J. was not a member.   † Died 1942.

physicists, but could also hope to leave Cambridge at the end of two years with the hall-mark of a Cambridge degree as the reward of their labours. The practical synchronisation of these two changes —the institution of research degrees at Cambridge, and the new regulations of the 1851 Exhibition Commissioners—undoubtedly brought to Cambridge from 1896 onwards many men who might otherwise have spent their years of research elsewhere, and it is probable that no one who was thus led to Cambridge and the Cavendish Laboratory had ever afterwards any reason to regret his choice.

In 1922, the 1851 Exhibition Commissioners published *Particulars of the Science Research Scholarships awarded by the Royal Commission for the Exhibition of* 1851 from 1891 to 1921 inclusive. From this publication it appears that down to and including 1895, twenty-nine physicists were elected as scholars, and of these only two went to Cambridge, both in the year 1895. From 1896 to 1921 inclusive, one hundred and three scholarships were awarded to physicists, and no less than sixty scholars of that number spent the whole or part of their time at Cambridge. In the photograph of Research Students in the Cavendish Laboratory in 1898, reproduced in J. J. Thomson's *Recollections*, out of the total number of sixteen students in the group, no less than nine were holders of these scholarships. These figures are a tribute to the renown of the Cavendish Professor of Physics, and to the wisdom of the University in introducing the new grade or class of Research Student.

The Research Student Movement was not popular in all quarters; for in several ways it tended to trespass on vested interests, or at any rate it was suspected of doing so. Some graduates thought that the degree which had cost them or their parents so much time and money was being given to these outsiders on cheap terms, and that this must tend to depreciate its value. Then again the new men from outside were competitors for the limited supply of good things in the way of studentships and fellowships, and also perhaps for other appointments within and without the University which men with the Cambridge hall-mark might hope to gain. Within the Cavendish Laboratory they were competitors for the limited facilities in the way of apparatus and the services of the workshop, and also (some may possibly have feared) for the attention and sympathy of the Professor. If any feeling of this latter kind existed it was certainly unjustified, for he had sympathies wide enough to include all who deserved them. Some members of the teaching staff, however, were inclined to show that they thought little of the newcomers, and this feeling found its way to the assistants, who were at first a little obstructive. But on the whole these mutterings soon passed away.

The part played by J.J. in incorporating Research Students into the Cambridge organisation is well worthy of notice. Research Students were given the status of B.A.—wearing the B.A. gown without the 'strings'. Many of those who worked in the Cavendish Laboratory were at Trinity, and they dined at the B.A. table, thus being projected as complete strangers into the midst of men who had spent the previous three years in close companionship. In such circumstances, it was quite natural that they should be regarded at first by some as intruders and 'not Cambridge men'. But J.J.'s action soon removed any feeling of this kind. From the outset he let all Cambridge know how he welcomed these newcomers, and he made the Research Students quick to realise that they were now 'Cambridge men' in the fullest sense and anxious to adapt themselves to Cambridge traditions and habits of life. And when one Research Student (Craig Henderson) became President of the Union, and another (Townsend) was elected to a Trinity College fellowship, it could be said that the new scheme was a complete success. But it was J.J. who brought this about, and the Research Students of those early years have had reason to be grateful to him for his constant watch over their interests. He in his turn found his reward in the number of young physicists who gathered round him year after year and played their part along with those trained wholly in Cambridge in producing the record of achievement which marked his tenure of the Cavendish Chair. The breaking down of all barriers, and the welding of all into one Cambridge School, was the work of J.J. He at first admitted the Research Students to the Laboratory without a fee, considering that to charge one would be a strain on their resources, but outside opinion was so strong against this that he was compelled to change his practice. The incident is of some interest, for it shows how far public opinion has travelled in such matters. It is now thought to be almost a matter of course that promising research workers should have an income found for them from public or semi-public funds, at least until they have had time to establish a reputation. On the other side, it must be remembered that the University found modest sums of money much harder to come by in those days than now.

It has already been explained that two separate circumstances— the opening of the University to post-graduate students from else-where, and the changed regulations for the 1851 Exhibition Scholarship—were contributory causes to the outburst of activity at the Cavendish Laboratory. It so happened that among the new arrivals there were several men of quite exceptional ability, and these combined with the Cambridge graduates formed a team such as can have seldom in the history of the world been found working at any scientific subject in the same building at the same time.

One of the new arrivals was Ernest Rutherford.* He had come from Canterbury University College, New Zealand, with an 1851 Exhibition Scholarship. While still at home he had read, as he told me, everything that J.J. had written, and had made up his mind that he was the man under whom he would like to work. Though Rutherford was at first afraid he had not made a good impression, J.J. in fact sensed his value immediately. He had come from New Zealand with a scheme of research ready prepared which led to his magnetic detector for electric waves. He got to work immediately on this, and in a few months it led to the detection of the waves over the distance of something like a mile and a half between the Cavendish Laboratory and the Observatory in Madingley Road. This was a much longer distance than had ever been achieved before—I think Lodge's experiments at Oxford (80 yards) held the record. The name of Marconi had not then been heard of. J.J. Thomson made enquiries in the City as to the possibilities of developing commercially a system of wireless communication based on this work. Those whom he consulted were of opinion that while the scheme would form excellent subject-matter for a

* The late Lord Rutherford, O.M., P.R.S.

prospectus, it was not likely ever to be of real commercial use. He was discouraged, and nothing further was attempted in this direction. Though not strictly relevant to the immediate topic, I will here introduce what J.J. said in a broadcast of 1934. 'As an illustration of the difficulty of forecasting the extent of the applications of a new discovery, I may mention that when a company was being formed for the supply of wireless, Lord Kelvin said to me that he thought the possibilities of its application might justify a capitalisation of £100,000, but certainly not more.'

Another valuable recruit arrived on the same day as Rutherford. This was J. S. Townsend,* who had gained distinction as a pupil of Fitzgerald at Trinity College, Dublin. He had aspired to be a fellow of that college, but under the antiquated system then in force, success depended on facility in answering examination conundrums in mathematics, and the successful candidates were so trained that they were familiar with every type of problem that had ever been set in any examination. Townsend did not think that to stay such a course would do his intellectual development any good, and came to Cambridge, where a more modern spirit prevailed. His first investigation was on the magnetic properties of solutions of iron salts and then he passed on to examine the properties of electrified gases, as we shall see presently.

Then there was J. A. McClelland,† another Irishman of powerful build and grave demeanour, friendly and capable. He came from Queen's College, Galway. So also did John Henry. Finally there was W. Craig Henderson,‡ who had been at Glasgow University under Lord Kelvin, and had served for a time as Lord Kelvin's scientific secretary. Mr Henderson has favoured me with some recollections for the purpose of this book, and part of what has been written above is due to him. He proceeds:

These four men were my constant companions while I remained at the Laboratory. Later our little group was increased by the addition

* Now Sir John Townsend, F.R.S., Wykeham Professor of Physics at Oxford.
† Afterwards F.R.S. and Professor of Physics at University College, Dublin. Died 1920.
‡ Now K.C. and leader of the Parliamentary Bar.

of Paul Langevin (from France),* John Zeleny from America† and Harold Wilson‡ from Yorkshire College, Leeds, and many happy evenings were spent together at the rooms of one or other of us. It was at one of these meetings that the idea was first raised of having a Cavendish Laboratory Dinner. It was at once acted on and the first dinner was held before the Christmas Vacation in 1897 and was so successful that this function became thereafter an annual event. There is an error in J.J.'s account of this first dinner. He says it was held in December 1898 at Bruvet's restaurant in Sidney Street. I have the menu card before me which shows that the date was December 9th, 1897 and the place of meeting was the Prince of Wales' Hotel. The only Toasts (apart from loyal Toasts) were 'Our Guests' proposed by Townsend, the reply being given by J.J., and 'Our old Universities' which, I see, I proposed. J.J. presided, and was as happy as a sand-boy, and at the Laboratory on the following day remarked that he had no idea that the Laboratory held such a nest of singing birds. In his 'Recollections' he mentions that our noisy gathering attracted the attention of the University Proctors and says that they did not enter the room 'being, I suppose impressed and, I have no doubt, mystified by the assurance of the landlord that it was a scientific gathering of Research Students'. This, however, was not the real reason. It was on learning that the presiding genius was the Cavendish Professor of Physics that they hastily withdrew.

Röntgen's discovery of the X-rays was published late in 1895, and was followed by a greater outburst of enthusiasm than any other experimental discovery before or since. Most physical laboratories had the means of taking X-ray photographs of hands, and this was tried on all sides: the next step was to give help to surgeons, who were not as yet equipped for taking advantage of the new discovery. Very many enthusiasts rushed into print on the subject, but for the most part this early effervescence was able to add little to what Röntgen had done: for he had worked out the more obvious questions raised by his original discovery pretty

---

* Langevin came from the École Normale Supérieure, with an introduction from Violle. Afterwards Professor at the Collège de France.
† Afterwards Professor of Physics and Chairman of the Physics Department, Yale University.
‡ Now Professor H.A. Wilson, F.R.S., of Rice Institute, Houston, Texas, U.S.A.

fully before he revealed anything to the world, or indeed, it is said, even to his own wife.

At the Cavendish Laboratory, the interest was naturally very great. Röntgen's paper was expounded by J.J. himself at the Cavendish Physical Society, the room being crowded to overflowing. A photograph was taken of the hand of one of the ladies present, and developed and shown during the lecture. In the early days efficient X-ray tubes were difficult to obtain commercially. Everett made and exhausted numerous tubes of the simple pattern then in use, in which a diffuse beam of cathode rays* was allowed to impinge on the wall of the glass tube. The X-rays spread out from the region of impact. Naturally tubes made in this way would not stand a heavy load. It was fancied that as the glass got less bright owing to fluorescent fatigue, the X-ray emission deteriorated. A little later, however, it was realised that the glass fluorescence had no necessary connection with production of X-rays, and metal targets came into use.

The tubes made by Everett were used by W.H. Hayles, the lecture assistant, who was the photographic expert of the laboratory. He took many photographs for surgeons practising at Addenbrooke's Hospital and elsewhere. Photographs were sometimes taken of bones which had been broken and reset in the past, and the resulting revelations were not always satisfactory either to patient or surgeon.

The following letter is interesting as showing the theoretical and experimental ideas in the earliest days.

J.J. Thomson *to* Professor Oliver Lodge:

6 *Scroope Terrace. Jan.* 19*th*, 1896.

I think the cause of the Röntgen photographs must be something not quite identical with the ordinary cathode rays as Röntgen finds that they are not affected by a magnet. I have been trying to get the photographs by putting the plate inside the vacuum tube so that the cathode rays may fall directly on a little ebonite box enclosing the plate and are so prevented from striking against the glass and causing phosphorescence. Under these circumstances I have never succeeded in getting the photographs. This looks as if phosphorescence as well as a cathode

---

* The meaning of this phrase is explained fully in Chapter IV

was necessary. That phosphorescence without a cathode is not sufficient is, I think, shown by some experiments I have tried with my bulbs without electrodes which though showing strong phosphorescence on the glass were quite inoperative as far as these photographs were concerned. It seems to me that both a cathode and phosphorescence are necessary.

On the whole I incline to the opinion that they are due to waves so short that the wave length is comparable to a molecule. Whether these waves are transverse or longitudinal is, I think, at present an open question. The absence of refraction is, I think, not surprising for these very short waves. If the excess of the specific inductive capacity over unity is due to the molecules setting themselves under the electric field we should, I think, expect it to be very small for these small waves. For supposing the half wave lengths were just equal to the length of the molecule, the force on the positive atom would be equal and parallel to the force on the — atom and there would not be a couple tending to make the molecule set, but merely a force tending to push it along.

I am trying to find whether there is any motion of the ether close to a cathode.

Scientific discussion turned a good deal on the relation between the discovery of Röntgen and the earlier work of Lenard. Some rather superficial commentators were inclined to take the view that Lenard in his experiments, by which he proved that the cathode rays could be got to penetrate into the open air through thin aluminium windows, had anticipated Röntgen, though Lenard himself made no such claim. Sir William Thomson and Sir George Stokes in particular were strongly opposed to this view.

Lensfield Cottage, Cambridge.
27th August, 1897.

Dear Professor Thomson,

For fear I should forget to mention it when we meet, I may as well put on paper a piece of information told me by the late Charles Brooke, F.R.S., as it might possibly be useful to you in preparing screens to shut out light while reducing to a minimum the obstruction offered by the screen to the X-rays.

You may remember that he devised apparatus for the self-registration, by means of photography, of magnetic and meteorological instruments. He told me he was much troubled to get an even black surface by means of lamp black. He tried laying it on by means of turpentine, but he could not get neat results. At last he found that the fault lay in the moisture, of which under ordinary circumstances lamp black contains a lot. He found that if he expelled the moisture by heating (I think it required going up to near a red heat) and then adding the turpentine, he got a mixture that he could lay on quite evenly and smoothly. I have not had occasion to use the thing, but apparently there would be no difficulty in preparing the mixture, which might be kept in a bottle for use as required. I should think thinnish paper thus varnished might be useful in exhibiting the compound nature of what Lenard got in atmospheric air outside his aluminium window, and tracing the behaviour of the two parts, utterly different in their nature, namely the X-rays given out from the first surface of the window and then traversing the window, and the cathode rays produced at the second face, and travelling in air till obstructed. Lenard took the phosphorescence produced outside as an indication of one and the same agent, and similarly as regards photographic effect, and so he missed Röntgen's great discovery.

Yours very truly, G. G. STOKES.

The discovery was soon made by J. J. Thomson that the rays caused the discharge of an electroscope. The known action of ultra-violet light in causing discharge of negative electricity from zinc made this a fairly obvious experiment to try, and in fact the discovery was made independently by several different workers elsewhere. But it was Thomson and his school who worked out in detail its true significance.

To explain what this significance is will require us to go back in time, and to refer to ideas often somewhat vague, which are scattered in the earlier literature about electric discharge.

In the first place, a gas is ordinarily a non-conductor of electricity; all experimenting on statical electricity (and after all, the study of electricity began with that) depends on this fact. If the air were a conductor, then obviously all electric charges on solid bodies would leak away through it. There are, however, limitations to this non-conductivity of air. If the electric tension to which it is exposed exceeds a certain limit, the insulation of air breaks down and a spark passes. If the air is moderately rarefied, this happens much more easily, and as the rarefication proceeds, the spark becomes more diffuse, finally broadening out into a band of light.

Evidently then a strong enough electric tension throws the air or other gases into a conducting state. There were other known methods of doing this. For example, a flame was known to be conducting from early times.

Now the kinetic theory of gases was firmly established at the time of which we are writing. According to this theory, the particles (molecules) of a gas are separated by spaces very large in comparison to their own dimensions. Since there are numerous difficulties in the way of assuming the empty interspaces to be electrically conducting, we are led to suppose that when a gas is in a conducting state the electricity must be carried across from a charged body to earth by means of charged particles, which move under the influence of the electric force. If, for example, we consider two opposed metallic plates, which are connected to the terminals of a battery, then positively charged particles or carriers in the gas will be attracted to the negative plate, and negatively charged carriers to the positive plate. The motion of these charges will constitute an electric current, and we can see that, given the production by some means of such charged carriers, the gas will become a conductor, in spite of the fact that its material is inherently discontinuous, with relatively large empty interspaces.

Ideas of this kind were scattered throughout the literature. They were considerably strengthened by the fact that electrolytic conduction through solutions was also believed to be of a convective nature. For example, in the electrolysis of dilute hydrochloric acid, while positive electricity in some way streams to the negative pole, hydrogen is also conveyed to the negative pole, and it seems very reasonable to assume that these two streams are not independent, but that they are different aspects of the same stream and that the hydrogen atoms actually convey the positive electricity. We notice further that, as Faraday discovered, there is a fixed relation between the amount of electricity conveyed and the amount of hydrogen liberated at the negative electrode. These facts receive their interpretation if we assume that each atom of hydrogen has a specific charge of electricity which it conveys with it through the liquid. Similarly, we assume that the negative carriers are chlorine atoms and that these too have a specific charge.

There are, however, very important differences between the conducting liquid and the conducting gas. The liquid is always ready to conduct,* and we must assume that the charged carriers are always present to a greater or less extent, in such a conducting liquid. (This is the hypothesis of electrolytic dissociation. Liquids such as a solution of sugar which are not conducting are held not to contain any free charged carriers.) On the other hand a gas in its ordinary state does not contain charged carriers, or at any rate it contains very few.

The charged carriers in electrolysis are called ions, and the production of them is called ionisation.† In an electrolytic liquid ionisation is a spontaneous process and the number of ions present at any moment represents the balance of this process and of the reverse process of the recombination of ions, positive with negative.

In a gas there is, at least on a rough general view of the matter, no process of ionisation spontaneously occurring.‡ In a flame or in the luminous discharge (spark at low pressure) there is a process of ionisation, and we shall see something of what this process consists in. In the meantime it is enough to notice that it must occur. Since the gas when removed from the flame recovers its insulating power, we must also assume that there is recombination. Ideas more or less of this general nature were put forward by various workers. The hypothesis of electrolytic dissociation in its modern form is due to Arrhenius. Giese put forward the hypothesis of ionisation in flames in 1882. Schuster in 1890 showed that it was possible to get the ions from the luminous discharge to diffuse through gauze into an adjoining space from which the discharge itself was excluded. In this adjoining space the air had lost its insulation, and a current could be passed by the use of small electromotive forces of a fraction of a volt. He did not, however, find it practicable to carry the investigations of gaseous ions further by this method. The use of flames, too, was unsuitable. All such

* Abstraction is here made of the phenomenon of electrolytic polarisation.
† This word is applied in medicine in a different way, and it will be best for the reader who only knows it in that connection to put the medical use out of mind altogether.
‡ We shall have occasion to refine on this statement later.

methods were complicated by the haunting suspicion that chemical processes occurring at the surface of the electrodes were in part responsible for the observed effects.

This hasty sketch is written rather to explain the general position of the subject in 1896 than to do strict historical justice to the earlier workers. In writing it, I have not been able altogether to avoid showing that wisdom which comes after the event. Although all that has been said could be justified by quotations from the earlier literature, yet it was obscured with a thick undergrowth of other suggestions, often coming from the same writers, which have not proved to be relevant. J.J. Thomson summarised the position as it appeared to him in 1893 in his *Recent Researches.** Anyone who glances at that summary will see what a flood of light was cast into dark places by the work which will now be described.

In the familiar cases of electrical conduction, as in the case of metallic wires, the current which passes increases proportionally to the electromotive force applied. If we apply double the voltage, we double the current, and so on without limit, so far as we can prevent the wire from getting hot, which alters its properties. One of Thomson's first experiments was to try how the current which can pass through air exposed to X-rays would be altered by varying the electromotive force, other conditions remaining the same. The current which could be passed was too small to be measured by a galvanometer, and it was necessary to have recourse to an electrometer, and observe the rate at which it charged up. An illuminating fact at once came to light. The current increased at first proportionately to the electromotive force, but then more slowly, until finally a stage was reached when it was independent of any further increase in the electromotive force.

At this point Thomson invited Rutherford, who had finished his work on the magnetic detector of electric waves, to come into partnership with him, and they worked out together many of the properties of the conducting gas in an epoch-making investigation. They found that the conducting power of the air persisted for a second or two after it had been blown away from the direct path of the rays, but that it soon reverted to the non-conducting

* P. 189.

state if left to itself. They found further that the very process of carrying a current deprived it of its conductivity. This was proved by blowing the air which had been exposed to X-rays through a tube containing an axial electrified wire. The air emerging from this tube lost its conductivity as soon as the central wire was electrified.*

This immediately explained the result before stated of the maximum value of the current, irrespective of increased electromotive force, for such a maximum must occur if the conducting power is destroyed by the very fact of conduction.

To put it in another way, the conducting particles are produced at a certain rate by the X-rays. If there were no cause tending to remove them they would increase indefinitely in number. If they are removed by the electric force which is acting, then the current which can be conveyed is evidently limited by the rate at which they are produced. This maximum current was called by Thomson and Rutherford the 'saturation current', presumably after the analogy of saturated solutions, for a solution of salt (e.g.) is said to be saturated when it is incapable of dissolving any more salt.

What, however, of currents which are *not* saturated? Or to take the extreme case, what happens if the current is evanescent, or if no current passes at all? What becomes of the carriers or ions in that case? Since the current-carrying capacity of the air quickly disappears if it is left to itself, the ions must disappear and there can be little difficulty in explaining how this happens. Each positive ion has an electrical attraction for a negative ion, and will unite with it when they come into contact. Thus the pair of ions cease to exist as such and contribute nothing to carrying a current.

There will now be no difficulty in understanding what happens in the intermediate cases. If a small electromotive force is applied,

---

* There is a curious slip at this point in the paper, and it is difficult to understand how it escaped two of the most acute minds of the age. The authors say: 'It is the current which destroys this [conducting] state, not the electric field; for if the central wire is enclosed in a glass tube so as to stop the current but maintain the electric field, the gas passes through with its conductivity unimpaired.' It is, however, illusory to suppose that the electric field would be maintained under these circumstances. It will be neutralised in the gas space, and thrown on to the glass tube, by a layer of ions which would accumulate on the outer surface of the latter.

the ions begin a slow procession, positive to the negative plate, and negative to the positive plate. This motion is resisted, and it soon comes to a limiting speed. A stone dropped from a moderate height continually gains in speed till it reaches the ground. A man hanging from an open parachute, whose motion is strongly resisted, quickly comes to a limiting speed, which does not change. This last case exemplifies the motion of the ions under a small electromotive force, the resistance dominating the situation. During the slow procession the ions recombine, and new ones are formed by the rays and take up the burden of carrying the current.

If now the electromotive force is increased, the ions take less time to go across and have less chance of combining. Hence they do more in the way of carrying current before their useful life is over: but they are still being generated as fast as before, so that clearly the current increases. The limit comes when the electromotive force is so great, and the time which ions take to reach the electrodes so small, that there is practically no opportunity for recombination to occur. The current is then 'saturated'.

Suppose now that, other things being unaltered, the plates are removed farther apart, so that a wider beam of X-rays can pass between them. There will now be more ions formed, and therefore, supposing them to be all used, the current will be larger than before. If an ample electromotive force was acting, the result of putting the plates farther apart should be to *increase the current*, which seems a paradoxical result when we compare the ionised gas with ordinary conductors. Thomson and Rutherford readily verified it, however, and this was felt to give a specially satisfactory check of the correctness of their point of view.

These were the main results of an epoch-making investigation. Rutherford gave me a copy of it one evening when I was in his rooms a few months later, a copy which I still have, and from which this account was written. He allowed me to see that he was proud of his share in it. It was read at the meeting of the British Association at Liverpool in 1896, and afterwards published in the *Philosophical Magazine*.

We have passed rather lightly over the technical difficulties of these experiments, but they were not small. The rays had to be kept at constant intensity to make the successive readings comparable, but this was far from easy, and indeed, at that time, there was no method available except incessant repetition, and sacrifice of all series of experiments which did not bear this test. Then again, the currents were too small to be measured with any available galvanometer. At this time very few laboratories had had much experience with quadrant electrometers, though these were described in textbooks, and it was rather naively assumed that if they were wanted, they could easily be made to work. J.J. himself wrote towards the end of his life: 'Another instrument which was exasperating to work with was the old quadrant electrometer. This not infrequently refused to hold its charge, and neither prayers nor imprecations would induce it to do so.'

Lord Kelvin was responsible for the most elaborate design of electrometer, and in these instruments the maintenance of the charge of the needle was satisfactory. They were costly and there was only one of them in the laboratory. Townsend used it. I rather think that J.J. had found that it defeated him. He used to say that he had been to a lecture by Lord Kelvin on contact electricity and that in the demonstration the electrometer invariably went the opposite way to what Lord Kelvin said it would. J.J. himself used what was called the Elliott pattern, after the name of the maker. I do not know who designed it, but (to plagiarise Oliver Heaviside in another connection) I suspect that it was primarily the Devil.

There was a shallow Leyden jar filled with sulphuric acid, which served both as the inner coating of the jar and as a 'dashpot' for damping the motion of the needle. I do not think any of us appreciated that a jar of soda glass or green bottle glass would not insulate well enough. Flint glass was necessary. But even with that it appears probable that the sulphuric acid soon crept over the surface and destroyed the insulation. Moreover, a sort of skin seemed on occasion to form on the acid, and when this happened the needle might come to rest anywhere. No modern electrometer uses sulphuric acid in this way, but in retrospect it is strange that the workers in the Cavendish Laboratory were content to struggle with it so long. We thought we ought to have been taught to use an electrometer in the demonstration classes, but when this was suggested one of the demonstrators asked: 'How many people do you suppose I could demonstrate to in a day if I did that?' In short, most of the workers in the laboratory had to admit that they could not master all the vagaries of the electrometer. J.J. was not regarded as exceptionally successful with it.

To return, however, to developments of the ionic thesis. The ions move under a potential gradient, and the steeper that gradient is, the faster they will move. The question arises what the speed actually is under a standard potential gradient of 1 volt per centimetre. The investigation of Thomson and Rutherford allowed of a rough general answer to this question. Suppose we have a current that approaches, but does not reach saturation. That means that the ions move across in a time which is comparable to the time required for recombination. For if they moved much faster than this there would be saturation: and if they moved much more slowly, there would be no approach to saturation. Recombination takes a fraction of a second, as may be shown by experiments in which the air is removed from the action of the rays by a blast, or by switching off the source of radiation. It could be concluded in this kind of way that the velocity must be of the order of 1 centimetre per second under 1 volt per centimetre. Rutherford now attacked the question more definitely, and devised various experiments to determine the velocity precisely. Since the time taken for the ions to move between the electrodes is only a fraction of a second, it is necessary to have time-measuring arrangements capable of dealing with these small intervals and the time of swing of a pendulum from rest through a determined arc was employed. Rutherford arranged so that the rays were shielded off from a space of some centimetres near one of two large plate electrodes, and he found how long it took for the ions to reach this electrode when the electromotive force was automatically switched on as the time-measuring pendulum passed a certain point.

The further details of the experiment are mainly of technical interest, and it is not necessary to go into them here.

Another worker in the laboratory, John Zeleny, went further

into this question of the velocity of ions, using methods whereby the velocity is compared directly with the velocity of a blast of air. The simplest form of the experiment is so to arrange that the ions have to move against the blast in order to reach the electrode. If the critical potential is so adjusted that they are just not able to do this, we conclude that the velocity under that potential is the same as the velocity of the blast. There are technical difficulties about this, and a method in which the ions move along the radius of a tube while the blast is along it proved more manageable. But the essential idea is the same. The interesting result emerged that the negative ions move more slowly than the positive. In dry air the velocity of the negative ions was found to be 1·87 centimetres per second, and that of the positive 1·36, the gradient being 1 volt per centimetre.

Although these investigations were not J. J. Thomson's own, it has been considered necessary to give an account of them, in order to fill in the picture. The various experimenters at the Cavendish Laboratory were to a great extent fitting in pieces of the same puzzle. The master inserted the main outlines. His pupils, with his help and encouragement, fitted in details. Some of them were more independent than others, and naturally their dependence tended to diminish as they grew older and more experienced. At times some of them were actually at issue with him. But, for all that, the picture would lack coherence if we only dwelt on those parts of it which he personally contributed. The plan adopted has been and will be to lay chief stress on these parts, but to fill in with briefer accounts of what was contributed by others. There were, of course, many workers in the laboratory whose efforts lay in other directions, not always less important: to deal with their work would lead us too far from the main theme.

CHAPTER IV

GREAT DAYS AT THE CAVENDISH LABORATORY (*continued*). CATHODE RAYS AND CORPUSCLES

WE have already had occasion to mention the cathode rays, a term which may not be familiar to the reader. By way of preface to what is to follow, we must now make the meaning of the expression quite clear. It originated in Germany (Kathoden-Strahlen).

Let us suppose the electric discharge of high tension to pass through a highly exhausted space, using as the negative electrode a flat metal disc. Then we shall see that an influence is propagated normally from this flat cathode in some respects reminiscent on a small scale of the beam from a searchlight: for its track is marked out by a blue luminous haze, and it produces a patch of light when it comes up against the wall of the glass tube, just as the searchlight does when it comes up against a cloud. If the cathode is concave the beam converges to a focus. If convex, it diverges. The direction is independent of the anode, which may be placed to one side, or may be a plate with a hole suitably placed through which the beam will pass. This beam or pencil constitutes the cathode rays. It differs from a searchlight beam in a striking particular, in that it can easily be bent about by bringing a magnet near it. If the magnet is so placed that the lines of force stretching between its poles are at right angles to the beam, then the beam will be displaced in a direction at right angles to itself and to the magnetic lines.

A further important fact about the cathode rays is that when they strike a solid obstacle, such as the wall of the glass tube, they heat it strongly when they impinge, and give rise at the same time to an emission of X-rays. These rays are not connected with the visible (green) fluorescent light which is given out by the glass wall, and which has already been mentioned. This is proved by the fact that if the cathode rays are received on a metal target, this gives out X-rays, but no visible light. Although, as we have mentioned, the path of the cathode rays, like the path of a search-light beam, is marked out by a luminous haze, this is not the essence of the phenomenon in either case. The track of the search-light, as seen from the side, is conspicuous if the air is misty: otherwise very much less so. Similarly, the track of the cathode rays is easily traced if the gas pressure is, say, ten-thousandth of the atmospheric pressure. If the pressure is much less, say a millionth of the atmospheric pressure, the track is no longer trace-able, but the characteristic effects are produced when the rays strike a solid obstacle.

There were two schools of thought about the nature of cathode rays. It may be said broadly that the English school considered them to be corpuscular, carrying a charge of electricity, and that the German school considered them to be of the nature of a wave propagation. In the present state of science this issue is by no means so definite as it was at the time between 1870 and say 1905 of which we are now writing. I do not think that it would be useful or would conduce to clearness if the difficulties of the pre-sent day, which were then undreamt of, were imported into the history of thought in those times. In the present chapter the issue will be presented in the way which appealed to contemporary thought, though in a later chapter something may be said about later discoveries which have reopened a question which seemed to be closed.

The corpuscular view was probably first insisted on by C. F. Varley in 1872, and Lord Kelvin was always strongly insistent on the merits of his contribution. The beautiful experiments of Crookes about 1879, which from the point of view of showman-ship have perhaps been scarcely rivalled in any field of scientific experiment, were also interpreted in this way, and although in some cases the interpretation was too naive, there can be no doubt that he strengthened the corpuscular view considerably.

Crookes was perhaps the first to show in a really clear and satis-factory form the magnetic deflection of the rays, though this had been in a sense foreshadowed by Plücker as early as 1858. The rays are found to be notably deflected by even weak magnetic forces such as can be produced by a small horseshoe permanent magnet. This experiment is quite easy to repeat, and has generally and rightly been considered crucial. We shall see why in the sequel. In the meantime, it is desirable to remark that a wire carrying an electric current experiences the same kind of force in a magnetic field, and is pushed at right angles to its own direction and to the magnetic force.

The most important exposition of the anti-corpuscular point of view was in a paper by Hertz in 1883. He made experiments with a view to detecting an effect of the cathode stream on a magnetic needle. These experiments seem open to obvious objections which he never mentions. As regards the magnetic deflection his point of view will best be explained in his own words (translated):

It seems to me probable that the analogy between the deflection of the cathode rays and the electromagnetic action is quite superficial. Without attempting any explanation for the present we may say that the magnet acts upon the medium, and that in the magnetised medium the cathode rays are not propagated in the same way as in the unmagnetised medium. This statement is in accordance with the above-mentioned facts and avoids the difficulties. It makes no comparison with the deflection of a wire carrying a current, but rather suggests an analogy with the rotation of the plane of polarisation in a magnetised medium.

Besides Hertz, E. Wiedemann and E. Goldstein, two other well-known workers on this subject, advocated a similar point of view. Von Helmholtz, who was Hertz's master, also seems to have supported it for a time, though, a few years later, he came to think otherwise.

Another important point tested by Hertz was to try whether the cathode rays, fired into a metallic vessel (known in this connection as Faraday's cylinder or Faraday's ice pail) would carry with them an electric charge, detectable by an electrometer connected with the vessel. He failed to observe this effect, but the design of his experiment was open to certain objections which were removed in a later investigation by Perrin in 1895, directed to the same question. Perrin got definite evidence that the rays carried a negative charge. J. J. Thomson, in a modification of

Perrin's experiment, showed that if the Faraday cylinder was put out of the line of fire of the cathode rays, it acquired a charge when, and only when, the cathode rays were so deflected by a magnet as to enter the cylinder.

J.J. Thomson previous to 1895 had been much exercised by these difficult and dubious questions, and he recapitulated the controversy in a course of lectures on electric discharge which he gave in 1894 and which I attended as a freshman. This was certainly not what a freshman ought to have been doing, but I do not regret it. Much of what has been said above recapitulates what I then learnt from him. He left the impression that he considered the magnetic deflection almost conclusive evidence for the corpuscular theory, and allowed us to see that he was not impressed by Hertz's suggestion of an analogy with the rotation of the plane of polarisation. The observation by Hertz that the cathode rays could get through gold leaf, and the development of it which had just been made by Lenard, who got the rays out into the open air through a thin aluminium window, was felt to be a hard nut to crack; it was difficult to envisage electrified particles as getting through an airtight metal partition, and Thomson was inclined, if I remember rightly, to think that perhaps a new corpuscular stream might be generated on the far side.

After the first excitement about the X-rays in 1895–6 was over, and after the more obvious points about X-ray ionisation had been worked out, Thomson returned to the study of cathode rays, with epoch-making results. This study was not directly connected with Röntgen's discovery, and might perhaps have been made earlier if that discovery had not turned his attention aside. We shall see later how these two branches of work ultimately coalesced, mutually fertilising one another.

Thomson had always been impressed by the magnetic deflection of the cathode rays, which distinguishes them so sharply from light and from X-rays, as giving the key to the whole problem. In this he differed from the German physicists, who, prepossessed with the other view, were inclined to emphasise the phenomena which seemed to them to confirm it. He began by measuring the amount of magnetic deflection. This he did by arranging that the beam of cathode rays should be immersed in a region of uniform magnetic force of known amount, produced by a large coil of wire. This field did not need to be very strong. It was found to be enough to use a field of 35 units, i.e. about 200 times the horizontal magnetic force of the earth, and this would bend the cathode rays used into a circle of 9 cm. radius. J.J., writing about this subject long afterwards, said:

> I had for a long time been convinced that these rays were charged particles, but it was some time before I had any suspicion that they were anything but charged atoms. My first doubts as to this being the case arose when I measured the deflection of the rays by a magnet, for this was far greater than I could account for by any hypothesis which seemed at all reasonable if the particles had a mass at all approaching that of the hydrogen atom, the smallest then known.

On the corpuscular view, the stream of charged particles constitutes an electric current, and experiences a lateral force in a magnetic field in the same way as a wire carrying an electric current. It is true that there is a difference, since in the one case it is a stationary piece of metal that experiences the force, and in the other (we suppose) a stream of separate electrified particles in rapid motion. This difference is partly bridged over by an experiment made by Rowland to which we shall refer again. But in discussing the deflection of the cathode rays it is assumed that their current-carrying aspect is the essential one. We calculate the sideways force as being the same as on an element (short length) of a current-carrying wire. In a magnetic field of 1 gauss the wire carrying 1 ampere experiences a force of $\frac{1}{10}$ dyne for every centimetre of its length. Now transfer this point of view to the cathode beam in a magnetic field of 1 gauss. If it conveyed a whole ampere, a whole coulomb of electricity would pass any point in a second, and 1 centimetre length of the beam would have a charge of $1/`v'$ coulombs, if 'v' were the speed. Upon this charge there would be a force of $\frac{1}{10}$ dyne. This illustrates by a special case how the sideways push on a moving charge can be calculated in terms of the velocity and the amount of the charge, and the magnetic field in which it moves.

Some writers have used with advantage the phrase 'magnetic

stiffness' of cathode rays to express the strength of the transverse magnetic force necessary to bend them, just as one measures the stiffness of a spring by the mechanical force necessary to bend it. (It must not be forgotten, however, that the spring is deflected in the direction of the transverse mechanical force applied, whereas the cathode rays are deflected at right angles to the direction of the transverse magnetic force applied to them.) Now the magnetic stiffness does not depend only on the mass of the particles, as the above quotation from Thomson might suggest, if the qualifying phrases were ignored. It depends really on two things, one of them being the velocity of the particles, and this as may be imagined is variable according to circumstances. But there is another quantity involved, which requires a little more explanation. Evidently the electric charge carried by the particles enters, since it is on this that the electromagnetic action depends, but the acceleration which the electromagnetic force produces in moving the particles sideways will be less if the particle is massive than if it is not massive. The mass of the particle and its charge enter into the question not independently, but as a ratio. It is fairly easy to see from another point of view that this must be the case. All the particles of the stream move along the same curved path. If we supposed two of them temporarily stuck together, nothing would be changed. The curved path would still be followed by the particles, now imagined to be Siamese twins instead of mere neighbours. The aggregate mass is double, but the charge is double also, and the change is without effect. This indicates that the mass and the charge enter as a ratio. This ratio combined with the velocity determine the amount of the magnetic deflection: or, conversely, knowing the magnetic deflection we can obtain some information about the ratio of charge to mass and the velocity. If we make a guess at the velocity, we can determine what ratio of mass to charge would follow. Thomson thought that the velocity was almost certainly large compared with ordinary molecular velocities.

It may be suggested that what we want to know is the mass, and not merely the ratio of mass to charge. As a matter of fact, the latter was the more instructive, because Thomson knew the ratio of mass to charge of ordinary atoms by the phenomena of electro-lysis, and it was a comparison of this kind that led him to doubt whether the cathode-ray particles could be atoms. He has not given us his provisional calculations in detail, but it is possible to reconstruct them. Take the case mentioned when the rays are bent into an arc of 9 cm. radius by a magnetic force of 35 units (gausses) applied transversely. If the rays consisted of charged hydrogen atoms, then, as in the electrolysis of water, about 10^{-5} grams of hydrogen are associated with a coulomb of electricity.* We can find what velocity a stream of such particles would need to have in order to bring up its magnetic stiffness to the observed value.† The required velocity is 31 kilometres per second. It may assist the imagination to recall that this is about the velocity of the earth in its orbit round the sun.

If the atom of hydrogen were left to find its natural velocity when in temperature equilibrium with the molecules of air in a room, the velocity would be about 2·6 kilometres per second, about $\frac{1}{12}$th the hypothetical velocity we have calculated for it in the cathode rays. Thomson did not think that 12 times the normal velocity they would have had anyhow was enough to confer upon hydrogen atoms the extraordinary properties possessed by the cathode stream: even allowing for the fact that they were electri-cally charged. The argument in this form was not conclusive. We are merely trying to imagine in a little more detail what Thomson hinted to us of his intellectual gropings at this stage.

It is not certain whether Thomson's first measurements of the magnetic deviation were made before or after his return from America in the autumn of 1896. At all events the experiments now to be described were made afterwards. Evidently it was necessary to know something more about the cathode rays than their magnetic stiffness if the argument as to their nature was to be made in any sense complete. One sufficient reason for this is that the magnetic stiffness is variable, becoming greater as

* The coulomb is 1 ampere-second.

† Thomson had earlier attempted to find out by a direct method, using a rotating mirror, what was the velocity of cathode rays. This is referred to in the letter to Schuster quoted on pp. 39–40. He afterwards withdrew the conclusion he had drawn from this attempt.

the tube is more highly exhausted and the discharge potential increased. It is not in itself a definite datum, because it involves not only the nature of the particles, but also their velocity. We must have some other information if we are to get any further. In mathematical language we require two equations to determine two unknown quantities.

Now there were other properties of the cathode rays which lent themselves fairly easily to measurement, and Thomson saw clearly that he was certain to get some result which would help to clarify his ideas if he measured (1) the heating effect of the rays, and (2) the electric charge carried. Both these effects, as we have seen, had been well and definitely observed in a qualitative way, and it could not be doubted that the measurement was feasible, and indeed comparatively straightforward to carry out. Neither of these quantities would be of any definite use *alone* because they would depend on the arbitrary intensity of the rays. It is no use (e.g.) cutting off a piece of rope at random, and expecting to find out anything of value by carefully measuring its length *only*. If, however, we measure the *length and the weight* of the same piece, we shall learn something about the character of the rope. In the present case it was necessary to measure the charge and the energy carried by a certain arbitrary quantity of cathode-ray stuff—no matter what quantity as long as we carry out both measurements on the same (arbitrary) portion.

Thomson now made a quantitative experiment on these lines. He measured the electric charge in the way already indicated, and by placing a thermo-couple inside the metal case or 'Faraday cylinder' into which the rays were received, he measured the rate of heating up of the thermo-couple as well as the rate of charging up of the Faraday cylinder and the electrical condenser connected to it. Knowing the thermal capacity of the one in calories per dyne, and the electrical capacity of the other in microfarads, he could compare the energy received with the electric charge received. But the energy depends on the measured velocity. In this way a relation is obtained between the electric charge on the one hand and the mass and velocity on the other: or, if we prefer so to express it, between the velocity on the one hand, and the ratio of charge to mass on the other. One such relation had already been obtained by measuring the magnetic stiffness. Combining the two relations it was possible to determine without ambiguity the velocity and the ratio of mass to charge. It is not useful to attempt to put into prose the simple algebraical process by which the individual values of these qualities can be disentangled from the pair of statements involving both. It must suffice to give the result.

Thomson had found that a field of 35 gausses would bend the rays with an arc of 9 cm. radius. His further experiments showed that the rays carried energy at the rate of $2 \cdot 6 \times 10^{10}$ ergs per coulomb of electricity. Combining these data, it is possible to deduce that:

The Velocity is 15,000 kilometres per second
and
The Ratio of Mass to Charge is 2×10^{-8} grammes per coulomb.

Thus the more complete information obtained by the measurement of charge and energy showed that, as Thomson had guessed, the velocity was enormously larger than molecular velocities, and that the particles were something entirely different from hydrogen atoms, having a much smaller ratio of mass to charge.

This conclusion, with the arguments which we have presented so far, was announced at a Friday evening lecture at the Royal Institution on April 30th, 1897. Thomson does not labour the momentous conclusion to which the experiments had led him, but says merely: 'These numbers seem to favour the hypothesis that the carriers of the charges are smaller than hydrogen atoms.'

It does not appear that this lecture made a great sensation in the scientific world, still less in the world outside. I do not think that I myself heard anything about it at the time, and only heard the conclusion he had reached some weeks later at Cambridge. The probabilities are that few of the audience really took in Thomson's argument which, after all, requires the assembling of a good many lines of reasoning which were not then familiar. However, they no doubt realised that he was saying that he had found bodies smaller than hydrogen atoms, a statement which, in the then condition of science, was thought to be paradoxical, or even self-contradictory, an atom being (it was said) the smallest portion of matter that did or could exist. He did not himself think that what he said made many converts, and he believed that some of his audience did not think he was speaking seriously.

Thomson had gone forward so far on a fairly secure path—the properties of the cathode rays which he had measured were qualitatively quite well established, and even conspicuous, and when once the conception of measuring them had been grasped, he was able to proceed so far without serious difficulty. But there remained a formidable obstacle in the path, and until it was resolved the whole position was uncertain, and might be found to rest on unsound basic hypotheses.

If the cathode stream really consisted of electrified particles, it ought to be capable of deflection by a transverse electrostatic force. Thus, if it passed between the plates of a condenser, the negatively charged particles should be attracted by the positive plate, and repelled by the negative. That this should be the case was clearly appreciated by Hertz in 1883, but he had not succeeded in making the experiment work, and he regarded its failure as telling against the corpuscular nature of the rays. Since Hertz's work, the matter had been carried somewhat further by Goldstein, who described an experiment, not difficult to carry out, which certainly seemed to show that under some conditions the rays could be electro-statically deviated. Goldstein's experiment consisted in arranging two wire cathodes along the length of a cylindrical glass tube. They were parallel to one another, and lay on either side of the axis. If only one of these was connected, then the other one acted merely as an ordinary shadow-throwing obstacle, and cast a shadow on the opposite wall, because it screened this wall from the cathode rays. But if the second wire was connected with the first, the shadow became very much broader, though at the same time less dark. That it became less dark was natural, because the second cathode was now a source of rays. But why was it wider? This might be explained by assuming that there was electrostatic repulsion by the second cathode regarded merely as an electrified body. The rays from the first cathode would curl away from it on either side and it would form a much wider shadow than before, when the rays were propagated in straight lines.

Though Goldstein's experiment was definite as far as it went, and a helpful contribution towards solving a puzzling problem, it was not felt to have fully clarified the situation. The conditions around an active cathode contained many elements of uncertainty. The electric force near a wire could not be uniform under any circumstances, and when a discharge was going, its value at any particular place was incalculable owing to lack of knowledge of the distribution of free electricity in the surrounding space. Moreover, a space which was traversed by a luminous discharge was subject to unknown conditions. It was a land of magic and mystery, where anything might happen. What was wanted was a deflection in a space where there was simply a uniform measured electrostatic force and no unknown complications or uncertainties.

Thomson had thought about these things a good deal, and he discusses the subject in his American lectures of 1896. He suspected that the conductivity of the residual gas might be the disturbing cause, and after his return with this clue in his mind he repeated Hertz's experiment, passing the beam of cathode rays between a pair of parallel plates, connected to a battery of storage cells. The beam was arranged to be narrow, and the position when it fell on the glass end of the tube could be pretty accurately located by the fluorescence.

Thomson at first got the same result as Hertz—no deflection where the battery was connected. But his attention being concentrated on the question of conductivity of the residual gas, he measured this in the usual way, and found it to diminish rapidly as the pressure in the discharge tube was diminished. This gave him encouragement to try whether he could observe the electrostatic deflection at the lowest pressures. It was found at a certain stage that the expected deflection occurred for a moment when the deflecting battery was connected, but that the fluorescent spot soon crept back to its undeflected position. Lowering the pressure still further, it was found that a permanent deflection could be obtained, proportional in amount to the voltage applied to the deflecting plates. He could detect it when this was only 2 volts. It was considered that the failure to get deflection at higher pressures was due to the accumulation of electric charges on or near the deflecting plates, which prevented the electric field between them being uniform and in effect protected the beam of cathode rays from really experiencing the lateral electric force which it was attempted to apply to it. This success in getting the electrostatic deflection greatly helped to clear the situation, and left little room for doubt that the corpuscular theory of the cathode rays was the right one. More than that, it formed the basis of an independent method of investigating quantitatively the properties of the rays, and checking the results already described.

Let us see what information can be got from observing both the magnetic and electric deviations, or arranging for a balance between them. It is easy to arrange that they shall give deflections in opposite directions, and we shall suppose this done. We shall suppose also that the electric and magnetic fields are uniform, sharply limited and coterminous, conditions which unfortunately cannot be accurately realised in practice: but for our purpose the simplification can be allowed. Let us suppose further that the electric and magnetic fields are kept fixed at a constant value. Then, if a particle travels very slowly along the length of the fields, it will be pushed sideways by the electric force in full strength, just as if it were not moving at all. But when we come to consider the magnetic force, the case is far otherwise. A slow procession of particles means a small electric current conveyed, and therefore the mechanical force on the procession or stream is small, and it only exerts a small sideways push on each of them. In this case then the electric deflection predominates.

Consider now the other extreme, when the stream of particles is moving very fast. In this case the push of the electric field is the same as before, but the push of the magnetic field is enormously increased, and if the motion is fast enough, it predominates and there is outstanding magnetic deflection.

It is clear that an intermediate speed must exist at which these two opposite deflections will neutralise one another. What this critical speed will be clearly depends on the ratio of the two fields.

If we doubled both, the balance would be preserved. If we increased (e.g.) the electric field only, the necessary speed would have to be increased in order that the magnetic push should still be able to balance the increased electric push.

We see then that the critical speed is tied up to the relative value of the fields. If we know the relative value of the fields, we can say, from the theory of electromagnetism, what speed would be critical. If the speed is unknown, we can fix it by determining the relative value of the fields which will make it critical. If, for example, we applied a transverse electric field of 100 volts per centimetre and found that we were able to compensate its action by a magnetic force of 10 gausses, the velocity must be 10^9 centimetres per second.* This was about what Thomson found in some of his experiments.

If the coterminous fields are 10 centimetres long, the particle will traverse them in 10^{-8} seconds. The theory which we have already sketched shows that the transverse force due to the magnetic field in the case mentioned is 10^9 dynes on every coulomb of charge conveyed. It would, of course, require an enormous number of particles to make up as much as 1 coulomb. However, that is not the essence of the matter. The calculations we are now considering apply to cathode-ray stuff in the aggregate, and are not limited to one particle, or to a thousand particles.

We wish now to consider what sideways drift this ought to produce on the stream when it traverses the full length of the magnetic field, the electric field being now removed, and the magnetic field acting alone. That depends on the mass associated with the coulomb of electricity. Let us make a tentative supposition about this, and suppose it were the same as in the case of hydrogen atoms in the electrolysis of water. We saw that this hypothesis broke down hopelessly before, but let us give it another chance, and see if it can do any better this time. A coulomb of electricity passing through acidulated water sets free 10^{-5} grams of hydrogen, very nearly. There are therefore 10^{-5} grams associated with a

* The system of electrical and magnetic units is of course so contrived that the only factors which enter are multiples of 10, except in cases when nature dictates otherwise.

coulomb of electricity. The question is then how far would a mass of 10^{-5} grams be moved by a force of 10^9 dynes acting for 10^{-8} of a second. Those who have studied (e.g.) the free fall of a stone under gravity will be able to answer this question. The answer is that the distance would be only $\frac{1}{20}$th of a millimetre, a distance in any case difficult to measure by the unaided eye, and imperceptible in the conditions of experiments like these.

Actually, however, when the magnetic field of 10 gausses alone acts, the sideways displacement in traversing this field is very conspicuous, and amounts to about 5 centimetres; so that evidently it would again be quite wrong to suppose, as we have done provisionally, that every coulomb of electricity was loaded with as much as 10^{-5} grams of mass, like hydrogen in electrolysis. The inertia of the stream is far less, and the distance it makes sideways is far more than could possibly be reconciled with this supposition, and in fact the large sideways displacement actually made shows that in the cathode-ray stuff there can only be about 10^{-8} grams associated with a coulomb of electricity or 10^8 coulombs associated with 1 gram. This confirmed the former result obtained by quite a different method, so that the position was now very much strengthened. In no case known before this time was electricity associated with so small a mass as in the case which we have cited: so that the provisional supposition which was proposed above was the best attempt that could be made to meet the facts. If we wanted to get the maximum of electricity on to the minimum of matter, charged hydrogen atoms as revealed in electrolysis represented the best that contemporary conceptions could do. But, as Thomson pointed out, this best was an entirely inadequate best. It was necessary to invent some kind of stuff such that a gram of it would carry not merely 10^5 coulombs, but 10^8 coulombs. In this preliminary account we have only used round numbers.

So far in discussing Thomson's work of 1897 we have avoided saying much about atoms or particles, because after all the phenomena described are not of such a nature as to reveal directly a particulate character in the cathode stream. They might equally occur if the matter concerned were a continuous fluid. But there could be little or no doubt that if the stream were to be regarded as material at all, it must consist of discrete particles. The low density of gases and their general behaviour had forced the conclusion that the molecules in them were moving freely, and were separated by wide interspaces, and the same must apply *a fortiori* to the cathode stream, which was much more tenuous still. Crookes had indeed spoken of the cathode stream as constituting a fourth or ultra-gaseous state of matter. What kind of particles did the stream consist of? Helmholtz had emphasised that the hydrogen atoms in electrolysis must be regarded as each carrying a specific charge. The following quotation is from his Faraday Lectures of 1881:

> The most startling result of Faraday's law is perhaps this. If we accept the hypothesis that the elementary substances are composed of atoms, we cannot avoid concluding that electricity also, positive as well as negative, is divided into definite elementary portions, which behave like atoms of electricity. As long as it moves about in the electrolytic liquid, each ion remains united with its electric equivalent or equivalents. At the surface of the electrodes decomposition can take place if there is sufficient electromotive force, and then the ions give off their electric charges and become electrically neutral.

If this conception of 'atoms of electricity' was to be retained it became very probable that the cathode-ray particles each carried its atom of electricity, that is to say, that it carried the same charge as the hydrogen atom or other monovalent atom in electrolysis. But, if so, it was necessary to assume that this charge was carried on a much smaller mass than the hydrogen atoms. This followed from the fact now proved that cathode-ray stuff could carry a coulomb of electricity on a much smaller mass than electrolytic hydrogen could do.

This then was the argument which led Thomson to the most important result of his scientific life—the existence of masses of a smaller order of magnitude than atoms. The cathode-ray stream was made up of these small masses, each charged with negative electricity. Thomson at this time called them 'corpuscles'.

It is a difficult matter to ensure the sufficient purity of gases used for experiments at these low pressures, but Thomson satisfied himself that the above properties of cathode-ray stuff were independent of the nature of the residual gas in the discharge tube, and also of the material of the electrodes. He summed up his further conclusion in the following way:

The explanation which seems to me to account in the most simple and straightforward way for the facts is founded on a view of the chemical elements which has been favourably entertained by many chemists: this view is that the atoms of the different chemical elements are different aggregations of atoms of the same kind. In the form in which this hypothesis was enunciated by Prout, the atoms of the different elements were hydrogen atoms; in this precise form the hypothesis is not tenable, but if we substitute for hydrogen some unknown primordial substance X, there is nothing known which is inconsistent with this hypothesis. . ..

Thus on this view we have in the cathode rays matter in a new state, a state in which the subdivision of matter is carried very much further than in the ordinary gaseous state: a state in which all matter—that is, matter derived from different sources such as hydrogen, oxygen, etc.— is one and the same kind; this matter being the substance from which all the chemical elements are built up.

We have explained above, in what seemed the simplest way, the main points of the investigation and the conclusion which it led to. But such simplified accounts, though they are useful to those who approach the subject for the first time, leave out a great deal that is of value and interest: and it is now necessary to amplify this preliminary statement in various directions without going too much into technical matters.

Hitherto we have given Thomson's results only in round numbers. The results as finally published in the *Philosophical Magazine* of October 1897 may now be given.

He found by the method of electrostatic deviation that the cathode-ray stuff weighed $1 \cdot 3 \times 10^{-8}$ grams for every coulomb of electricity conveyed. The other method gave $0 \cdot 85 \times 10^{-8}$ grams per coulomb. It cannot candidly be said that these measurements are in close agreement; for each showed a difference of 20 per cent from the mean of the two, an uncertainty not considered tolerable in any measurement made for everyday purposes, such as say the weighing out of groceries. Thomson, however, was not seriously discomposed by this. He had applied quantitative methods where none had been successfully applied before, and rightly thought that if he had not been on the correct scent, much wilder discrepancies than these would have revealed themselves.

Thus his methods of measurement brought into quantitative relation the current carried by or with the cathode rays, the heating effect when they impinged on a solid, the electrostatic and the magnetic deflections, and the driving potential difference. On the other hand there were, undetermined and at disposal, the velocity and the ratio of charge to mass of the particles. There were more equations than unknown quantities: and if the whole conception of cathode rays as charged particles had been wrong (and a little earlier distinguished scientific workers were arguing against it) no kind of reconciliation of these equations would have been possible. Thomson rightly judged that a rough reconciliation could be provisionally accepted, and he rightly relied upon the conclusion to which it led.

The story is often told that Newton laid aside the calculation in which he connected the moon's motion with terrestrial gravity because there was a discrepancy of the order of 10 per cent due to a wrong value of the earth's radius. This story is not universally received, and I for one have never been able to accept it: but whatever Newton may have done, it is very certain that Thomson would not have abandoned his thesis even temporarily in like circumstances.

It has been said that an important new discovery is always pronounced (1) not new, and (2) not true. I do not think that this was conspicuous in the case of Thomson's discovery. At the same time it would be a mistake to suppose that he was writing on an entirely clean sheet when in 1897 he investigated the properties of the cathode rays in a quantitative manner. The quantitative theory of their motion in a magnetic field had been discussed by Stokes, and by Schuster. Hertz had shown by an incidental remark in his paper of 1883 that he understood very well that, if the corpuscular theory of the cathode rays were accepted, and if the electrostatic deflection could be observed, this could be combined with the magnetic deflection to give the velocity of the particles; but, being unable to get the electrostatic deflection, he had no occasion to pursue the matter further. Schuster in 1890 had come nearest to anticipating Thomson. He clearly recognised that the mass associated with 1 coulomb of electricity in the cathode rays could in principle be found from the magnetic deflection and potential difference. The value of this achievement was recognised by the Royal Society, who, partly on this ground, awarded to him the Copley Medal, their highest honour, in 1931. In attempting to apply the method Schuster was less successful. He over-estimated the effect of the resistance of the residual gas in the tube in retarding the particles, but his experiments seem to have been subject to some other source of error which is difficult to trace, and which is not here of primary concern. At all events he concluded that the cathode particles were charged atoms. No one before Thomson had suspected their true nature.

Thomson's method of measuring the charge by an electrometer, and the energy by a thermo-couple, seems at first sight very different from the method (originally used by Schuster in 1890) of measuring the potential difference between a cathode and anode, of which more will be said below. But the difference is not really fundamental except as regards experimental technique. For how, after all, is the potential defined? It is the work done on a unit quantity of electricity in passing from cathode to anode. Thomson's measurements amount to determining calorimetrically the energy associated with that quantity of the stuff which carries unit charge: and this energy is acquired in passing between cathode and anode. Thomson's method measures the energy actually imparted to the rays. Schuster's measures the energy they ought to acquire, assuming that they start at the cathode, and then acquire the full possible amount. Thomson thought the assumption was not justified in fact, and he explained in this way some not inconsiderable discrepancy between his own measurement and those made subsequently by others who used the potential method. However, later work has confirmed the latter as the more accurate.

In the summer of 1897 J.J. was bubbling over with enthusiasm over his work on cathode rays. The first I heard of it was from himself. I was at the time only an undergraduate, but he knew I think from the questions I had asked him after his lectures that I was as

eager to hear as he could be to talk, and chancing to meet on King's
Parade he began to unfold to me what he had been doing—telling
me that the cathode rays had now 'turned out' to be particles, and
particles quite different from atoms. My rooms were at that time
in Whewell's Court, but I did not want to interrupt the tale he was
unfolding by stopping there, and walked on with him past St
John's and the Round Church to the other entrance of Whewell's
Court in Sidney Street, where he left me, after standing talking
for a few minutes. I hope I may be forgiven for recalling this
trivial incident, but perhaps it will help to illustrate the reasons
why his pupils were devoted to him. Unfortunately, I cannot fix
exactly when this occurred. It was probably in the summer term
of 1897, and it is rather difficult to understand why he did not
find plenty of more experienced listeners. I do not think he could
have found one more appreciative.

It was probably a week or two later that he gave an account
before the 'Cavendish Physical Society'. In this case the meeting
assembled without a very clear idea of what was coming, and there
was loud applause when the Professor explained his methods and
wrote on the blackboard the simple calculations which showed
what could be got from them. He showed us his discharge tubes.
'But', he said, 'I must apologise for not having them exhausted.'
The truth is that this was quite a serious undertaking in those days.
The only method of exhaustion available was a Töpler mercury
pump worked up and down by hand. This apparatus was by no
means portable and even when it was in position it had to be
worked for half a day before a good cathode-ray vacuum was produced, particularly if the apparatus contained large metallic electrodes which gave off gas interminably. As Thomson remarked
in a review of those days, a new meaning was found for the dictum
that 'Nature abhors a vacuum'.

It is of course entirely normal that pioneers should have to encounter this sort of difficulty—the reason being that it does not
occur to anyone to do the work of making a smooth and straight
road to the goal until someone has shown that there is something
to be found when you get there. It is like climbing an Alpine peak.
The first to succeed have to get up without the chains or other

helps which make it comparatively easy for their successors. The modern airpumps which make high vacuum work comparatively easy were developed mainly because the work of pioneers like Crookes, Röntgen, J.J. Thomson and Lenard had shown how rich this field of research could be.

We may here say something about the word electron, which is now used for the cathode-ray particles. It is commonly stated that J.J. Thomson discovered the electron by his researches on the cathode rays in 1897. It may therefore surprise many readers to learn that the word was proposed by Dr G.J. Johnstone Stoney in a paper published by the Royal Dublin Society in 1891. Stoney's suggestion was adopted by Larmor in 1894, and the word was taken into use by Lorentz and others, and soon came into fairly general use.

Stoney proceeded on the view that a monovalent atom in electrolysis was charged, and gave up its charge when free hydrogen was liberated. He calculated the amount of this charge on the basis of the (not very definite) knowledge of the number of hydrogen atoms in a cubic centimetre of the gas which is obtainable from the kinetic theory of gases, and he proposed the name electron for it; at the same time emphasising the importance of this quantity as a natural unit of electricity. This point of view clearly leads to the conception of a charge of electricity, not attached to an atom, and therefore having a relatively negligible mass. Accordingly when Thomson detected such charges with a relatively negligible mass in the cathode rays, writers adopted the word electron for them. Thomson, however, did not do so for many years. He continued to use his original word corpuscle. During the war period he wrote little or nothing on the subject, but in his post-war writings he ceased to speak of corpuscles, and conformed to what is now the universal custom.

We may now take the opportunity here of mentioning Thomson's work on the question of electromagnetic mass, which was of much earlier date than the time at which we have now arrived. In 1881, he investigated from a theoretical point of view the inertia of a moving electrified sphere of radius a, carrying the charge e, say. This sets up magnetic force in the surrounding space, and since this magnetic field represents energy, it follows that the motion of the sphere implies that energy is required to set it going, and create the magnetic field. This is another way of saying that it possesses mass, and, with certain assumptions, these ideas can be made quantitative, and the mass related to the radius of the sphere and the charge it carries. In his paper of 1897 Thomson did not seem to favour the view that the mass of his corpuscles was mainly explicable in this way, though the basic idea had been his own. Later, this view became fashionable by the advocacy of Lodge and others.

It is not difficult to see that if the mass is expressible in this way at all, it must be of the form e^2/a with some numerical coefficient, because if any other form were proposed, e/a for example, then the unit of length would be involved, and the quotient could not express a mass. The difficulty is in finding the numerical coefficient. This is likely on general grounds to be of the order of magnitude unity; Thomson, by a long calculation, gave originally the value $\frac{4}{15}$th, but it was afterwards amended by himself and others. The whole theory, however, apparently comes to grief over the mechanical properties which have to be attributed to keep it spherical, and it is now generally abandoned.

CHAPTER V

CLOUDY CONDENSATION AND THE CHARGE CARRIED BY ELECTRONS

THE atomic theory had been built up primarily from philosophical and then from chemical considerations, without observing any effects which could be attributed to atoms individually as opposed to their statistical behaviour in large crowds: and indeed, it was natural that this was so, for it is only with difficulty, and by the close pursuit of recondite observations made under very special conditions, that we can examine any effect attributable to individual atoms, so as to be able to say 'there goes one' and 'there goes another'. Naturally, observations made in the mass are much more difficult to interpret than observations made on individuals. It would be like interpreting history with reference only to the behaviour of large bodies of men, excluding all knowledge of personalities. It can be imagined that under these circumstances the study would become very difficult, and the conclusiveness of interpretations open to a great deal of doubt.

The observations of the Cavendish school which we have described so far were all of this statistical nature: for example, the cathode-ray corpuscles were examined by methods which made no attempt to distinguish individuals, and afforded no possibility of counting the number involved. The possibility of observing the effects of individual atoms or electrons depends in most cases on their being electrically charged. Thus we can observe a cloud drop found round an individual charged atom or electron. We can observe the crowd of ions generated by a single electron when they are suitably multiplied as in the Geiger Counter; or more simply, we can observe under a magnifier the scintillation produced when one rapidly moving charged atom hits a phosphorescent screen of zinc sulphide. An intermediate class of cases is in the Brownian Movement, when we begin to observe the effects of a moderate number only of atoms, not electrically charged. In cases of this kind we do not *see* individual atoms, except in a general and extended sense of what is meant by seeing. To go back to our historical comparison, the atom which we 'see' is, among atoms, rather like a (successful) Guy Fawkes among men. He releases effects which extend to an orbit much beyond his own person, and definitely and conspicuously enough to allow one to infer his own individual existence and action even if we do not see him individually.

We now come to the era when atomic effects of this kind first began to be realised, and we shall show how they gradually led to much more definite knowledge about the electron as an individual.

We begin with an account of the work of C. T. R. Wilson,[*] at that time a young Cambridge graduate. His long and distinguished scientific career has shown a singular tenacity of purpose, for by far the greater part of it has been devoted to the study of clouds. He was familiarly known in the Cavendish Laboratory as 'Cloud Wilson' to distinguish him from H. A. Wilson,[†] who was working there at the same time. His first publication dated back as far as 1895, before the discovery of X-rays, and it is remarkable how his work, begun in a totally different connection, was helped and fertilised by Röntgen's discovery, when it came.

Aitken had shown that the ordinary production of clouds, e.g. by slight rarefication of moist air in a flask, was conditional on the presence of dust particles which act as nuclei of condensation, and when these had been got rid of by repeated small expansions, which brought them down, further condensation was not readily obtained, the air remaining supersaturated instead of relieving itself by the formation of cloud. Now this pretty obviously raises the question of how far we can go without producing condensation in the absence of nuclei. It would seem that if the expansion and consequently the supersaturation is carried far enough, condensation must occur. The problem had been attempted by others before Wilson, but without conclusive results, and we shall not enter on the discussion of this.

[*] Afterwards Jacksonian Professor at Cambridge and C.H.
[†] Some of H. A. Wilson's work also had reference to clouds as we shall see.

We shall quote at some length (though with abbreviations) from Wilson's Nobel Lecture of 1927, to show how his attention had been directed to the subject and how his investigations developed. He says:

In September 1894 I spent a few weeks in the observatory which then existed on the summit of Ben Nevis, the highest of the Scottish hills. The wonderful optical phenomena shown when the sun shone on the clouds surrounding the hill top, and especially the coloured rings surrounding the sun (coronas) or surrounding the shadow cast by the hill top or observer on mist or cloud (glories) greatly excited my interest, and made me wish to imitate them in the laboratory.

At the beginning of 1895, I made some experiments for this purpose—making clouds by expansion of moist air after the manner of Coulier, and Aitken. Almost immediately I came across something which promised to be of more interest than the optical phenomena which I had intended to study. Moist air which had been freed from Aitken's dust particles, so that no cloud was produced even when a considerable degree of supersaturation was produced by expansion, did appear to give a cloud if the expansion and consequent supersaturation exceeded a certain limit. A quantitative expansion apparatus was therefore made in which given samples of moist air could repeatedly be allowed to expand suddenly without danger of contamination, and in which the increase of volume to be made could be adjusted at will.

It was found that there was a definite critical value for the expansion ratio ($V_2/V_1 = 1\cdot25$) corresponding to an approximately fourfold supersaturation. In moist air which had been freed from Aitken's nuclei by repeatedly forming a cloud and allowing the drops to settle, no drops were formed unless the expansion exceeded this limit, while if it were exceeded, a shower of drops was seen to fall. The number of drops in the shower showed no diminution however often the process of producing the shower and allowing the drops to fall was repeated. It was evident then that the nuclei were always being regenerated in the air.

Further experiments showed that there was a second critical expansion corresponding to an approximately eightfold supersaturation of the vapour. With expansions exceeding this limit dense clouds were formed in dust-free air.

While the obvious explanation of the dense clouds when the second supersaturation limit was exceeded was that we had here condensation occurring in the absence of any nuclei other than the molecules of the vapour or gas—those responsible for the rainlike condensation which occurred when the supersaturation lay between the two limits from the first excited my interest. The very fact that their number was so limited and yet that they were always being regenerated, together with the fact that the supersaturation required indicated a magnitude not greatly exceeding molecular dimensions, at once suggested that we had a means of making visible and counting certain individual molecules or atoms which were at the moment in some exceptional condition. Could they be electrically charged atoms or ions?

In the autumn of 1895 came the news of Röntgen's great discovery. At the beginning of 1896 J.J. Thomson was investigating the conductivity of air exposed to the new rays, and I had the opportunity of using an X-ray tube of the primitive form then used which had been made by Prof. Thomson's assistant, Mr Everett, in the Cavendish Laboratory. I can well recall my delight when I found at the first trial that while no drops were formed on expansion of the cloud chamber when exposed to X-rays if the expansion were less than 1·25, a fog which took many minutes to fall was produced when the expansion lay between the rainlike and the cloudlike limits; X-rays thus produced in large numbers nuclei of the same kind as were always being produced in very small numbers in the air within the cloud chamber.

J.J. Thomson had himself already made a contribution to this subject in his *Applications of Dynamics to Physics and Chemistry* published in 1888. He pointed out that the effect of electrification on a drop would be to retard evaporation, while the effect of diminishing size is to promote evaporation, because of the greater vapour pressure over a convex surface; the effect of electrification would be the opposite, and it would tend to retard evaporation, because of the greater energy of the electric charge when concentrated on a drop of diminished size. He showed how these considerations could be compared quantitatively.

Now evaporation may be regarded as the process inimical to the formation of cloud drops. The question is whether the incipient drop will evaporate into nothingness or not. We see then that Thomson's earlier work had shown that there was a reason why charged particles should act as nuclei of condensation when without the charge they would fail to do so. He advanced further along this line of thought in a paper of 1893, 'On the Effect of Electrification and of Chemical Action on a Steam Jet, etc.', which had a marked influence on Wilson's trend of thought. Wilson applied methods of calculation akin to those of Thomson to his own observations of the rainlike condensation on X-ray ions, and had even used them to deduce a good value for the ionic charge, but he did not publish this result because Lord Kelvin on a visit to the Cavendish Laboratory had urged him strongly to keep theory out of his paper. We may remark in passing that this was very characteristic of Lord Kelvin, who was enthusiastic about other people's experiments, but, at any rate in his mature years, was apt to consider their theories deplorable.

Parallel with this investigation by C. T. R. Wilson was an investigation by J. S. Townsend, who, as we have seen, was one of the first arrivals at the Cavendish Laboratory under the scheme for advanced students. After some preliminary work in the field of magnetism, he started an investigation on electrified gases, and observed that the gases liberated by electrolysis carried an electric charge, and that they were capable of causing cloudy condensation when brought into contact with moist air, even when it was not saturated. There was every reason to assume that the charged particles acted as centres of cloudy condensation. These clouds were observed to subside at a measurable rate, the flat surface of the cloud being tolerably well defined. The fall of these minute drops, like the fall of a parachutist, is entirely dominated by the resistance of the air, and is not at all comparable to the freely accelerated fall of a stone. Townsend pointed out that the rate of subsidence gave the means of estimating the mass of the droplets by a calculation due to Stokes, who had found the terminal velocity of a small sphere of known density and diameter falling through a gas of known viscosity. Combining this with the total mass of the cloud, the number of droplets could be deduced, and hence (assuming that each had its share of the charge) the fraction of the total measured charge that was carried by each. This step made by Townsend was of great historical importance in the development of the subject, though many writers have ignored it.* J. J. Thomson, however, makes due reference to it in his *Recollections and Reflections*. Townsend had a familiarity with hydrodynamics which was probably not general among the workers in the laboratory, and the experimental hint afforded by the definite

* See for instance Lamb's *Hydrodynamics*, 4th ed. p. 595 footnote.

rate of subsidence of the cloud no doubt suggested to him at once the applicability of Stokes' calculations. But reference to Stokes' original paper 'On the Effect of the Internal Friction of Fluids on the Motion of Pendulums'* shows that he himself had not overlooked the application to the droplets in clouds. He had the ordinary clouds of the atmosphere in mind. He concludes that 'It appears that the apparent suspension of the clouds is mainly due to the internal friction of the air'.

Stokes himself was not far off—he was often to be seen in the Cavendish Laboratory at that time, and I remember helping him to find a piece of apparatus belonging to him which happened to be in the very room in which Townsend was working—but it does not appear that the investigation owed anything to his personal inspiration. It was based on what he had written forty-seven years before.

Townsend's results were published in February 1897 and were of considerable interest in themselves. He found that the droplets in his cloud each carried a charge of 10^{-19} coulomb, but he did not attempt to draw any far-reaching conclusions from this result. He said, however, that there was nothing in it to disprove that the carriers have the actual atomic charges. On the other hand, it is possible to produce clouds in a somewhat similar way having uncharged nuclei, as Townsend himself showed a little later in an investigation of nuclei produced by ozone; so that the general bearing of the results was not altogether clear. It was possible that some of the nuclei in the electrolytic cloud were uncharged; and if so the specific charge of those which *were* charged would be larger and undetermined in amount.

The method invented and used by Townsend is not the only one by which the number of drops in a cloud can be found. Other methods have already been glimpsed in the quotation from C. T. R. Wilson which has been given. Aitken in his dust counter had allowed the drops to fall on a glass plate, and had counted them under the microscope. Barus used diffraction rings, and Wilson and J. J. Thomson also contemplated this method. But Thomson,

* *Trans. Camb. Phil. Soc.* Vol. IX, p. (8), Dec. 9th, 1850. Reprinted in *Mathematical and Physical Papers*, Vol. III, p. 1.

as we shall see, made successful use of Townsend's method in establishing the electronic charge, and historically this method is the practical and successful one.

There will now be no particular difficulty in understanding the fundamentally important determination of e, the charge on a gaseous ion produced by X-rays. This was carried out by J.J. Thomson himself, on the basis of the investigations which have just been described, and the work done by Rutherford on the velocity of the ions.

The general idea was to expose a known volume of air uniformly to the X-rays, and when the steady condition had been reached, to determine (1) the number of ions present—using the subsiding cloud, (2) the current under a known small potential gradient.

These data, combined with a knowledge of the velocity of the ions under the same potential gradient, gave Thomson the means of determining e, the ionic charge.

In the case contemplated, there will be a procession of ions towards the boundaries, positive ions migrating towards the nega-tive electrode and negative ions towards the positive one. The current near the positive electrode will be carried almost entirely by negative ions, and that near the negative electrode by positive ones. The current will be the same across any cross-section, and midway between the electrodes both kinds of ions take part in conveying it. We shall for simplicity of exposition suppose that the positive ions move at the same speed as the negative, though this is only roughly true. To show how the ionic charge is found from the data, it will be easiest to take a concrete case of one of J.J. Thomson's experiments.

He found the total number of drops to be 4×10^4 per cubic centimetre. There were therefore 2×10^4 positive ions and 2×10^4 negative ions. Consider first the positive stream. The density of positive electricity in the stream is $2 \times 10^4 \, e$, if e is the charge on each ion. The voltage applied was two Leclanché cells, i.e. 3 volts, and the electrodes were 2 cm. apart. This gives a potential gradient of 1·5 volts per centimetre, and according to Rutherford's data, obtained earlier, the velocity of the ions under this gradient would be 2·4 cm. per second. This is the velocity of the stream of positive electricity (e.g.) past a cross-section in the middle of the vessel. The amount of positive electricity carried past such a section would be $2 \cdot 4 \times 2 \times 10^4 e$ or $4 \cdot 8 \times 10^4 e$, when e is the charge on an ion. An equal amount would be carried by the negative stream in the opposite direction.

The electric current was measured and found to be $1 \cdot 035 \times 10^{-14}$ amperes for every square centimetre of cross-section of the conducting space, so that the quantity of electricity conveyed would be $2 \cdot 07 \times 10^{-14}$ coulombs per second; and half of this, or $1 \cdot 035$ coulombs, would be carried by the positive ions.

We conclude then that the quantity $4 \cdot 8 \times 10^4 e$ is $1 \cdot 035 \times 10^{-14}$ coulombs, or that e is $2 \cdot 15 \times 10^{-19}$ coulombs.

This was Thomson's result, though the method of presentation has been altered in a way that may perhaps make rather less demands on the reader's scientific knowledge. It need scarcely be said that the small currents had to be observed by means of the quadrant electrometer, using a known electrostatic capacity. They were far too small to be detected by the most sensitive galvanometer.

Thomson and Rutherford, in their investigation of the ions from X-rays, had found that these ions were completely removed by passing through a filter of glass wool. When the subject was still relatively unexplored anything was possible, and it seemed natural to interpret this result as proving that the ions were so large that, like smoke or dust particles, they could not get through the interstices of the filter. J. S. Townsend, however, who often discussed the subject with them, was not inclined to accept this point of view, but thought that the effect was rather due to the small size of the ions than to their large size. He pictured the ions as diffusing to the side of the narrow channel through which they had to pass, and sticking there. In order to test this view, he wished to make the experiment more definite and metrical, and substituted for the gauze an assemblage of small parallel brass tubes through which a stream of ionised air of known velocity could be made to pass, so that the loss of ions passing through it could be observed by conductivity measurements: the geometrical conditions being thus made definite, he was able by a rather complicated reduction of the experiments to deduce the coefficient of diffusion of the ions through the gas, air or other gas in which they were contained.

Motion of ions through the air can be set up in either of two ways. They can, as it were, be selectively taken hold of by the electric field, which lets the molecules of the surrounding air alone: or they can be driven along by diffusion which will occur when there is a greater partial pressure due to ions in one stratum than in the next, diffusion being the tendency to drive the ions through the gas until the partial pressure is equalised. The motive force in this case acts independently of the air, but the motion is obstructed in the same way in both cases by the resistance of the air through which the ions are moving, which may be thought of as 'passive' resistance. Evidently something should come out of comparing the effectiveness of these two ways of setting the ions in motion under the same resistance. Reasoning on these lines Townsend showed in 1899 how the velocity in a unit electric field could be brought into relation with the coefficient of diffusion. He showed that a knowledge of these two quantities would allow us to fix the product of the ionic charge and the number of molecules in a cubic centimetre of any gas—hydrogen or other. This product specifies of course a certain quantity of electricity. It can be compared with the quantity of electricity required to set free half a cubic centimetre of hydrogen in electrolysis, and is found to be the same. There is here a proof of the identity of the ionic charge and the charge carried by the hydrogen ion in electrolysis. This, it will be observed, does not depend upon the counting of cloud drops, or indeed on any independent observation of individual ions, and it does not fix the absolute value of the charge, independently of a knowledge of the number of molecules in a cubic centimetre. The early estimates of this quantity made by Maxwell from the viscosity of gases were only moderately satisfactory, and could not do much more than fix the order of magnitude. Nevertheless, Townsend's result, by showing the identity of the ionic charge in gases and the atomic charge in electrolysis, gave an invaluable strengthening of the whole position which was being arrived at in the Cavendish Laboratory.

Townsend had thus shown that the charge on the ions formed in air by X-rays was the same as that associated with the atom of hydrogen in electrolysis, and had considerably strengthened the suspicion that this charge would prove to be a kind of natural unit; hence there was plausibility in suggesting that this was also the charge carried by the particles or corpuscles in the cathode rays. But, as we have seen, this was a question on which very large issues depended, for if such were the specific charge of the cathode particles, it would follow that the mass of these particles was of the order of one-thousandth part of the mass of a hydrogen atom: and the merely presumptive argument was hardly enough to carry so momentous a conclusion.

The mass of the particles was one unknown quantity, and the charge was another. There could be no real satisfaction until both could be specified in absolute units. The magnetic deflection methods could give the ratio of mass to charge. The cloud methods could give the charge. The difficulty was to apply both these methods to the same kind of particle. The magnetic and electrostatic deflections could be applied to the cathode rays, but it was impracticable to apply the cloud method to these rays, because they could only be obtained in a high vacuum and the cloud method, involving the use of an atmosphere supersaturated with water vapour, clearly could not be applied in such conditions.

The position at which Thomson had now arrived, and the way out, will be understood from the following quotation:*

In a former paper (*Phil. Mag.* Oct. 1897) I gave a determination of the value of the ratio of the mass, m, of the ion to its charge, e, in the case of the stream of negative electrification which constitutes the cathode rays. The results of this determination, which are in substantial agreement with those subsequently obtained by Lenard and Kaufmann, show that the value of this ratio is very much less than that of the corresponding ratio in electrolysis of solutions of acids and salts, and that it is independent of the gas through which the discharge passes and of the nature of the electrodes. In these experiments it was only the value of m/e which was determined, and not the values of m and e separately. It was thus possible that the smallness of the ratio might

* *Philosophical Magazine*, December 1899.

be due to e being greater than the value of the charge carried by the ion in electrolysis rather than to the mass m being very much smaller. Though there were reasons for thinking that the charge e was not greatly different from the electrolytic one, and that we had here to deal with masses smaller than the atom, yet, as these reasons were somewhat indirect, I desired if possible to get a direct measurement of either m or e as well as of m/e. In the case of cathode rays I did not see my way to do this; but another case, where negative electricity is carried by charged particles (i.e. when a negatively electrified metal plate in a gas at low pressure is illumined by ultra-violet light), seemed more hopeful, as in this case we can determine the value of e by the method I previously employed to determine the value of the charge carried by the ions produced by Röntgen-ray radiation (*Phil. Mag.* Dec. 1898). The following paper contains an account of measurements of m/e and e for the negative electrification discharged by ultra-violet light, and also of m/e for the negative electrification produced by an incandescent carbon filament in an atmosphere of hydrogen. I may be allowed to anticipate the description of these experiments by saying that they lead to the result that the value of m/e in the case of the ultra-violet light, and also in that of the carbon filament, is the same as for the cathode rays; and that in the case of the ultra-violet light, e is the same in magnitude as the charge carried by the hydrogen atom in the electrolysis of solutions. In this case, therefore, we have clear proof that the ions have a very much smaller mass than ordinary atoms; so that in the convection of negative electricity at low pressures we have something smaller even than the atom, something which involves the splitting up of the atom, inasmuch as we have taken from it a part, though only a small one, of its mass.

It is necessary now to explain how this next step was taken. The discharge of negative electrification from metals by ultra-violet light had been the subject of several interesting pioneer investigations. Hertz in the course of his investigations on electric waves had required to observe minute sparks in his receiver, and these were difficult to see when they were unshielded from strong light. He therefore boxed up the receiving spark gap, but this was found to produce an unfavourable effect. This was ultimately traced to the ultra-violet light from the transmitting spark, which fell on the spark gap of the receiver and made the passage of this spark easier. He found further that the cathode surface of the spark gap was the sensitive spot. Hallwachs later found that the same effect could be greatly simplified and improved by avoiding the spark altogether, and merely using a clean piece of zinc attached to a gold-leaf electroscope. This would lose a negative charge under the influence of ultra-violet light from an electric arc, or a piece of burning magnesium ribbon. The further study of this effect was made by Elster and Geitel, who eventually developed the photoelectric cell, the basis of the talking film industry and of many other commercial applications. J. J. Thomson followed all these investigations closely, and he was struck by the dissymmetry between positive and negative electricity which they revealed. If, for example, the piece of clean zinc used in Hallwachs' experiment is positively electrified, not the slightest discharge is produced by ultra-violet illumination. This dissymmetry was not of course a new thing—the appearance of oxygen at the anode and hydrogen at the cathode in electrolysis was an obvious example of such dissymmetry. Nevertheless, I think that Thomson was struck with the way in which negative electricity would come off from the cathode, both in the cathode rays and in the photoelectric effect, without any easily observed parallel effects of positive electricity from the anode.

Another point which had impressed him was that Elster and Geitel had found in 1890 that at low gaseous pressures the discharge of negative electricity from zinc or other electropositive metals was checked by a magnetic field applied so that the lines of force were parallel to the surface, and thus at right angles to the lines of electric force. Elster and Geitel had only speculated vaguely on the interpretation of this observation, and Thomson too, in describing it in his *Recent Researches* in 1893, had mentioned it as an empirical fact, without attempting any interpretation. But if I remember rightly, he made some reference to it in 1897, when he was giving at the Cavendish Laboratory a verbal account of his work on the cathode rays, and said that though Elster and Geitel had not given any quantitative data on the magnetic and electrostatic fields used in their experiments, yet, if reasonable guesses were made as to the value of these, it would seem that the carriers of the negative electrification away from the zinc must be more like the cathode-ray 'corpuscle' than like charged atoms, or than the disintegrated zinc dust which Lenard and others had believed to be concerned.

In the meantime he had determined the charge on the Röntgen ray ions, and saw that it should be possible to make similar determinations on the carriers set free by ultra-violet light acting on clean zinc, as well as to determine their path under the simultaneous action of electrostatic and magnetic forces.

When it became known in the laboratory that he was going to attempt this, great interest was felt, and one prominent worker remarked to me that he thought it was 'tempting providence' to try this experiment. I repeated this remark to the Professor, when I knew the successful result, but he seemed not quite to see the fun of it and said that he had felt pretty confident.

In the form of experiment used by Elster and Geitel, there was an electrostatic force driving away negative particles from the illuminated zinc surface, and, simultaneously, a magnetic force was applied at right angles to the electric force. This had the effect of bending the path of the corpuscles away from the straight, and naturally the discussion of the magnetic deflection is more complicated than in the experimental arrangement used for cathode rays. For in that case the electric force gives the particles a velocity, and then acts no more; the magnetic force is applied farther on outside the region of the electric force. In the present case the electric force continues to accelerate the particle in the initial direction, while the magnetic force bends it away from that direction. It must suffice to mention the result of analysis, which shows that the particle traverses a cycloidal path. Naturally, the curvature is greater the stronger the magnetic force which bends the particle away from its original direction. The curvature is less the stronger the electric force which urges it in that original direction. If the exciting ultra-violet light is admitted through an insulated wire gauze placed parallel to the plate of zinc, then, in the absence of magnetic force, the charged corpuscles will be driven straight to this gauze by the electric force. As the magnetic force is increased their paths will be bent, but if the gauze is large, this bending will not at first prevent their reaching it. Ultimately, however, the curvature will become so great that the corpuscles are bent right back and can never reach the gauze. The reader may be reminded that a cycloid is generated by a circle rolling on a straight line. A point on the circle describes the cycloid. A familiar image is afforded by a speck of mud on a cartwheel, which traces out a cycloid in space. The speck of mud can never get farther away from the road than the diameter of the wheel. Similarly with the corpuscle. Increase of the electric force increases the diameter of the generating circle. Increase of the magnetic force acts more powerfully still in contracting it. It is a question of the ratio of the electric force to the square of the magnetic force. If this ratio is adjusted so as to make the diameter of the rolling circle just small enough to prevent the electrified particles reaching the gauze, we shall have the data required to determine e/m for the particles.

In practice the gauze was at a fixed distance from the plate, the magnetic force was maintained at a fixed value, and the electric force (applied voltage) was diminished until it just ceased to make any difference to the current whether the magnet was on or off. This voltage was taken to be the critical one. If the conditions had been ideal the current ought to have suddenly fallen from full value to zero when the magnetic force was constant and the electric force passed up through the critical value. This was very far from being the case, and the method is by no means a good one, if a very accurate value of m/e is the main object in view. But at the time of which we are writing this was not at all the primary object. Thomson was exploring quite virgin territory. It is even possible that at the time his experiments were begun he did not feel sure that the effect of the magnetic field discovered by Elster and Geitel, but not interpreted by them, was at all that contemplated in the theory we have outlined above. 'If this view of the action of the magnetic field is correct...', he wrote. But there can be no doubt of his sound judgment in proceeding on these lines even if a very good result was scarcely to be hoped for. He had read Elster and Geitel's work and knew that some kind of measurement could be obtained in this way: the main and urgent question was not so much whether the corpuscles liberated by ultra-violet light had a value of m/e exactly the same as those of the cathode rays, but whether or not they had a value of a different order of magnitude from that of atoms. Thomson wisely limited himself to making Elster and Geitel's experiment quantitative, and did not spend time in looking for a better method. No doubt in the end he could have found one. But there were other territories waiting to be conquered, and it was not well to delay too long over this one.

It may be remarked that this method does not differ fundamentally from the method of dealing with the cathode rays originally tried by Schuster in 1892, and afterwards carried out in greater perfection by Kaufmann and others. It consists in determining the potential differences applied to give the cathode rays their velocity, and also the amount of the magnetic deflection. Experimentally, it is simpler to let the electric and magnetic forces act simultaneously. Theoretically, it is somewhat of a complication, because the electrostatic force continues to act on the corpuscle after the path has become oblique to this force, or even at right angles to it. Thus electrostatic deflection occurs, and tends to oppose the magnetic deflection. We may therefore consider this method as being in theory a rather complex compromise in spite of its experimental simplicity.

Having got the value e/m the next point was to determine e, the specific charge of the corpuscles which carry the negative charge away from zinc illuminated by ultra-violet light. The great advantage of these corpuscles was that it was possible to work with them either in a high vacuum or in a gaseous atmosphere. The high vacuum conditions were necessary for the part of the experiment already described, because it was only so that the motion was free from interference by collision with molecules of the surrounding atmosphere and only so that the calculation of the cycloidal path has any validity. It is found in fact by experiment that the magnet has no appreciable effect at atmospheric pressure. To determine the charge by the cloud method, it is necessary to work at atmospheric pressure, and it was further found necessary to have an electromotive force acting when the cloud was formed, since otherwise the corpuscles remained near the zinc, and no cloud was formed in the body of the chamber.

The general method of performing the experiment was the same as that used in the case of the X-ray ions, and it is not in accordance with the plan of this book to enter into minor details of the work. The result was to give a value of e of $2 \cdot 3 \times 10^{-19}$ coulombs, practically the value found for the X-ray ions.

Thomson next proceeded to investigate another effect which had been noticed by Elster and Geitel. It had been known for a long time that red-hot bodies were not under all circumstances able to retain an electric charge. The earliest experiments of this kind were made by DuFay in 1733, and several other eighteenth-century experimentalists had occupied themselves with it. The first person who recognised that positive and negative electricity behaved differently was Frederick Guthrie, who is now best known as the leader in founding the Physical Society. His work was followed up in more detail by Elster and Geitel. We cannot here dwell in detail on any of their early work, except in so far as is directly necessary. Elster and Geitel discovered conditions under which negative electricity was discharged freely, while positive electricity was not discharged at all—just as in the case of the illuminated zinc plate. This result was obtained by using a red-hot filament of carbon, which had been well freed from gas by keeping it hot, and which was used in an atmosphere of hydrogen or in a high vacuum. They found further that the negative discharge could be checked by a transverse magnetic field.

These facts were enough to put Thomson on the track. It was clear that under the conditions negatively charged particles could be given off by the filament, and that these could be driven away from it by an electromotive force. He suspected, by a rough computation, that they were the same as the particles given off by zinc, under the action of ultra-violet light. The experiment was made quantitative in substantially the same way as in the previous case. The carbon filament was arranged very near a metal plate, with another metal plate opposed to it: the potential difference between the plates was increased until the magnetic force failed to diminish the electric current. The resulting value of e/m was nearly the same as for the experiment with ultra-violet light and zinc.

Let us now sum up the results which have been obtained so far. e/m for cathode rays 5×10^7 coulombs per gram of stuff.

For ultra-violet light corpuscles 7.3×10^7 coulombs per gram of stuff.

For incandescent carbon corpuscles 8.6×10^7 coulombs per gram of stuff.

As regards the actual charges it was found that

e for Röntgen ray ions was 2.2×10^{-19} coulombs.

e for ultra-violet light corpuscles 2.3×10^{-19} coulombs.

Thomson concluded that the three former kinds of particles were all of identical nature, being characterised by the same value of e/m. The identity of values is not very accurate, but, if we remember that the value was about a thousand times as great as for the atoms of matter, we shall see that the variation was not much to trouble about at this stage of the investigation.*

Of the three kinds of particles, he had only been able to determine the actual charge carried in the case of ultra-violet light, and in this case the charge was the same as for X-ray ions in air. But, having satisfied himself that the three classes of particles were the same, he necessarily attributed the same specific charge to the other two. Townsend's results had shown definitely that it was the same as the charge on the hydrogen ion in electrolysis.

In this way the evidence for the electron was made complete. The results were communicated to the British Association at Dover in the autumn of 1899 under the title 'On the existence of masses smaller than the atoms'. This occasion was a joint meeting with the French Association for the Advancement of Science at Boulogne. The available records of what was said in the discussion afterwards are rather disappointing; but it is remembered that some members of the chemical section, led by H. E. Armstrong, were dissatisfied. It seems that J.J., in his oral exposition, introduced some speculative views about the revival of Prout's hypothesis in a modified form, and perhaps also his notions about the periodic law, based on Mayer's experiment with floating magnets

---

* The modern values, which are the result of many converging lines of evidence, are $e/m = 1.76 \times 10^8$ coulombs per gram and $e = 1.59 \times 10^{-19}$ coulombs.

(see p. 139). (see p. 139). The chemists perhaps felt the difficulty of accepting these negatively charged bodies as the universal particles of which everything was made, having regard to the obvious difficulty that ordinary matter was electrically neutral. They were probably disposed to regard the integrity of the atom as the Ark of the Covenant which must not be touched: and indeed if we consider how hardly this position had been won in the face of the pretensions of the old alchemists, it is possible to feel some sympathy with their point of view. Armstrong spoke with great vigour and earnestness, dramatically unfolding a roll of black calico (representing a blackboard) covered with chemical symbols; but his arguments do not seem to have been recorded and I have not been able to discover what they were. He was accustomed to say about this time that J.J. and his school were insufficiently instructed in chemistry and that but for this defect in their education they would have known better; and something to this effect may have been expressed or implied. A.W. Rücker probably voiced the general feeling better when he said that in Prof. Thomson's paper the section had given to the French visitors of its best. Oliver Lodge, who read the next paper, said characteristically that he had been so interested and excited by the previous communication that he found a difficulty in collecting his thoughts sufficiently to expound his own!

It was considered by J.J. himself that this was the occasion when his views really made headway outside the small circle immediately surrounding him at Cambridge.

CHAPTER VI

MORE ABOUT IONS AND ELECTRONS

Townsend, having completed his work on the diffusion of ions, looked round for another subject of research, and began experimenting with a view to detecting the charge carried by the rays from radioactive bodies. This was to be detected by means of the quadrant electrometer. A vacuum had to be made to prevent the charge leaking away through the ionised gas. Tests were made to see how far this had succeeded, by testing the conductivity of the residual gas—for in those days the modern methods of making a high vacuum were not known. I was working in the same room as Townsend about this time, and I saw the first stages of the investigation carried out. I was then absent for a time owing to illness. When I came back in January 1899, I asked Townsend if he had succeeded in detecting the charge from radium—he said 'No, but I have got something better than that out of it'. He then explained to me that he had found that at low pressures the current which would pass through the ionised gas could exceed the 'saturation' value if the electromotive force was sufficiently increased. For small electromotive forces the current increased proportionally; for larger electromotive forces it remained constant, or saturated, being over a certain range independent of the electromotive force. For still larger electromotive forces, it increased further and rapidly. This last effect he interpreted as follows. A negative ion when moving fast enough collided violently with a neutral molecule, and ionised it, the original ions thus breeding new ones, and causing a rapid increase in the current. He also mentioned some experiments made by Stoletow in 1890 on the photoelectric effect. Stoletow had found that the current between the illuminated zinc plate and another plate opposite to it increased with the distance, the electric force being maintained the same. At least, Stoletow's results implicitly contained the evidence for this statement, though he had not brought it out explicitly. Having no theory to guide him, he was not able to select the most telling conclusions from his numerous experiments as Townsend was able to do.

J.J., who came in every day, did not at first accept Townsend's view of these effects, and urged an alternative explanation of Stoletow's results, based on a theory of electrical double surface layers at the zinc surface, which, he urged, were broken up by the action of ultra-violet light, liberating negative electricity. This theory had been explained in his American lectures. Townsend, while not denying that it might explain some of Stoletow's results, said that he did not see how it could explain the increase of current with distance, under a constant electric force: because, according to this view, the whole generation of ions would take place very near the cathode. This particular result of Stoletow's work is not explicitly mentioned in J.J.'s American lectures, and I do not think it had previously been present in his mind.

Townsend in the meantime himself set up an apparatus in which a zinc plate was illuminated by ultra-violet light. The advantage of this was that the negative ions (electrons in this case) started from a definite place. It was possible to determine quantitatively the number of ions produced by a negative ion travelling 1 centimetre, at various gas pressures and electric forces. J.J.'s opposition, although I heard it expressed almost daily for a week or two, gradually melted away, and he became convinced; no doubt quite forgetting his original attitude.

Townsend had taken a most important step in showing that comparatively slow motion of ions, produced by weak electric fields, could produce secondary ionisation in this way. The earlier experiments of Stoletow indeed contained implicit evidence of this. But no one before Townsend had given the interpretation, and to consider Stoletow as a prior discoverer is entirely beside the mark.

J.J. himself in his paper of 1897 had experimented in detail on ionisation by cathode rays, and had been led thereby to understand and overcome the difficulty of observing electrostatic deflection of the rays (see p. 86). Cathode rays are moving negative ions, according to the view expounded in that paper, and this was accordingly a clear case of ionisation by collision. These facts were no doubt present in J.J.'s mind, and he saw how helpful the conception of ionisation by collision could be in explaining the origin of the ionisation in the electric discharge. On February 5th, 1900, he read a paper before the Cambridge Philosophical Society under the title 'Ionisation of Gases in the Electric Field'. A title alone can scarcely give valid priority and it was unfortunate that J.J. should later have cited the date of reading of the paper without explaining where any contemporary account of its contents could be found—for there is none in the *Proceedings* of the Cambridge Philosophical Society. However, the omission can fortunately be supplied. Mr S. Matthews, the librarian of the Society, has unearthed the following in the *Cambridge University Reporter* of February 13th, 1900, which was no doubt supplied by J.J. himself at the time of reading.

Prof. *J.J. Thomson. Ionisation of Gases in the Electric Field.*

The view put forward in this paper is that the Ionisation of a gas in an electric field is brought about by the presence of ions already in the field. These ions move under the electric force and acquire energy which can be spent in ionising the gas. It can be shown that this view would explain why an electric field of definite strength is required to produce discharge, why a thin layer of gas is electrically stronger than a thick one, why the electric strength diminishes with the pressure of the gas until a critical pressure is reached when the strength is a minimum, as well as many phenomena connected with the discharge through gases at low pressures.

J.J. published a paper in the *Philosophical Magazine* of September 1900, under the slightly altered title of 'Genesis of Ions in the Electric Discharge through Gases', with a footnote to say that it had been read before the Cambridge Philosophical Society on February 5th. The contents were generally the same as those indicated in the short abstract above cited: no quantitative theory was developed, and it was assumed that secondary ionisation by positive ions occurred in the gas space near the cathode a view which Townsend writing shortly afterwards did not accept.

It has become rather widely known that Townsend complained

that his ideas had been used in this paper without proper acknowledgement. I think this was the case, but only to a slight extent. Townsend did not claim that the idea of ionisation by collision was new. Indeed, he himself wrote (*Nature*, August 9th, 1900):

Recent researches have shown that gases are rendered conductors of electricity when negatively charged ions move through them with a high velocity. Thus the cathode rays and Lenard rays possess the property of ionising gases through which they pass. (See J. J. Thomson, *The discharge of electricity through gases*.) Becquerel has also recently shown that the conductivity produced by radium is due to small negatively charged particles given out by the radioactive substance. In these cases the charged particles which ionise the gas move with nearly the velocity of light.

Some experiments which I have recently made show that the ions which are produced in air by the action of Röntgen rays will produce other ions when they move through the gas with a velocity which is small compared with the velocity of light.

J.J. in his paper of February 5th and of September 1900 assumed this last as a working hypothesis, and referred to the earlier experiments of Stoletow in confirmation of it. That Stoletow's work carried this implication was pointed out by Townsend to J.J. and myself in a triangular conversation, and J.J. had clearly forgotten this when he drafted the paper of September 1900. When he did draft it is not on record; it was presumably not drafted when it was 'read', i.e. orally expounded; for if it had been drafted then, it would have been handed in for publication. Townsend's own work was first published in *Nature* of August 9th, 1900, and must, I think, be considered the prior publication. He here makes the first published reference to Stoletow in connection with this problem.

Such, to the best of my recollection and belief, is the true history of these events. As regards the explanation, a passage in a letter from G. F. C. Searle, a lifelong colleague and associate of J.J., is illuminating. Dr Searle is not writing with special reference to this incident, but generally. He says:

J.J.'s mind worked in strange ways. He could not always remember how an idea had got into his mind. I did not meet this phenomenon much myself, but it caused difficulties to a good many people. He would be told by someone or would read somewhere some new idea. Later on he would find the idea floating in his mind and he would suppose that the idea was original to himself and would treat it as if it were. This gave the appearance that he was claiming as his own ideas which others had already published. I am sure he was unconscious that he was using the work of others.

Other correspondents too have independently written to me in the same sense. J.J. would repeat to them as if it was a new idea what they had told him a day or two before. This, however, was in private. In public he was in general most careful to treat the work published by a pupil, as that pupil's exclusive property. Indeed, some think that he went too far. I have heard it said that there were cases when a pupil, having got a considerable post on the basis of his published work, failed to justify it afterwards. It was even said that at one time there was a prejudice in America against Cavendish men on this account, but I cannot vouch for this, and do not know any names. It is certain at all events that a great many chairs in America were filled by J.J.'s pupils.

The theory of ionisation by collision was most illuminating, and to a great extent supplied the link which was required to connect up the researches made on the self-supporting discharge, and the other class of researches specially developed in the Cavendish Laboratory, when the ions were supplied by an external agent, such as X-rays or the rays from radioactive bodies. It will, however, probably occur to the reader that one point remains rather nebulous. Granting that the mechanism of collision could breed fresh ions from any that might already be there, what reason was there for supposing that there *were* any there? Perhaps it would not be too fanciful to compare this with the difficulties met with in early days of the bacterial theory of disease. Granting that bacteria could multiply in a nutrient medium, where did they come from? Pasteur and others showed that a few of them were present in the air. Similarly C.T.R. Wilson at the Cavendish Laboratory showed by admirably designed experiments that a few ions were always present in the air. These experiments were published in 1900. If ions are normally present in the air, it ought to be a conductor under moderate electromotive forces, and most people having a knowledge of experimental electricity would have said that there was no evidence of anything of the sort. If it were true, for example, a gold-leaf electroscope would not retain its charge. But, after all, did it retain its charge? No, not indefinitely; but that was presumedly due to failure of the insulating support. There was, however, some evidence that this was not the whole story, but it was conflicting. Dust might be a complication, and the matter had not really been adequately explored. Perhaps many thought that so minute a residual effect was hardly worth troubling about.

C. T. R. Wilson had, however, always obtained a few drops in his cloud experiments at the degree of expansion which, in the presence of X-ray ions, gave abundant drops: and the inference seemed to be that a few ions were always present. However, he was anxious to prove the fact by electrical methods. He conceived the simple idea that a gold-leaf electroscope could only lose its charge along the insulating support if there was a fall of potential along the insulator. If the far end of the insulator was kept at the same potential as the leaves, or a higher potential, then a collapse of the leaves could not possibly be due to this cause, and must be due to leakage through the air. In this way he was able to prove that there was a constant production of ions in the air, at the rate of about 14 of each sign per c.c. per second.

These researches were of fundamental importance. Not only did they show that ions were always available for starting the electric discharge, but they developed in other directions. It is from them that the discovery of the cosmic rays derives.

Before leaving the history of this period, it will be well for the sake of completeness to deal with some other researches which really attach themselves to it. We have seen how J.J. originally determined the charge on the X-ray ion, using the method of a subsiding cloud. He now undertook a revision of this.

To SIR OLIVER LODGE:
Cavendish Laboratory, November 25th, 1902.
Dear Sir Oliver,

The value of $6 \cdot 5 \times 10^{-8}$ is a misprint for $6 \cdot 5 \times 10^{-10}$ [electrostatic units]. I think this value is too large. I have just finished a new deter-

mination of it, using the same method as before, i.e. depositing the
cloud on the ions, but using several devices to make the *rate* of expansion greater: the cloud begins to deposit on the negative ions before
it does so on the positive ones, hence if the rate of expansion is not
very rapid the drops round the negative get an appreciable size before
the expansion reaches the value required to give a fog round the positive ions, the big negative drops form convenient centres of condensation and the moisture collects round them rather than round the positive,
thus the number of drops is little more than the number of negative
ions and not the number of negative ions + the number of positive.
In my old apparatus I was catching little more than the negative drops,
so that the number of ions was more nearly double the number of drops
than equal to it. In my new apparatus I can prove I am catching the
positive as well as the negative, for the number of drops with an expansion theoretically big enough to catch both the + and the − is twice
the number with an expansion big enough to catch the − only. The
new value of e is about $3 \cdot 5 \times 10^{-10}$. I think this value gives the most
reliable way of finding the number of molecules in a c.c. of gas at
760 and 0° C. I forget the exact number it gives but it is not far from
4×10^{19} which agrees well with the value got by other methods.

Yours ever, J.J. THOMSON.

About 1903, H.A. Wilson thought of a somewhat different
method, using an electrified cloud containing negative ions only.
It was got by a carefully regulated expansion, as explained in the
first part of the above letter.

This method occurred to H.A. Wilson while reading one of
Townsend's papers. Some mention was made of the force on a
charged droplet in a field, and he thought of using a vertical field
to move the drops upwards, and balance the weight which causes
them to fall. The rate of fall in the absence of the electric field determines the diameter of the drop; and hence its mass, and the force
exercised by gravity upon it. If we neutralise this, we get the
potential gradient required to exercise this same force electrically,
and hence the value of the electric charge.

So far we have spoken as if one individual drop was being observed, and in the later development of the method by R.A.
Millikan in America this was done. H.A. Wilson, however, by
carefully regulating the expansion used to produce a cloud, got a
cloud containing negative ions only, and observed the rate of

subsidence of the top of this cloud, with and without the electric field. The cloud itself naturally subsides at the same rate as the drops comprising it.

The great advantage of this method is that it is not necessary to estimate either the number of drops in the cloud or the number of ions present at the moment of its formation.

The result obtained was that the ionic charge was $1 \cdot 0 \times 10^{-19}$ coulombs. This differs appreciably from Thomson's original value of $2 \cdot 2 \times 10^{-19}$ coulombs. Thomson had, however, before this revised his determination with improvements and located certain sources of error. The revised result was $1 \cdot 3 \times 10^{-19}$ coulombs and this was in fair agreement with H.A. Wilson's. The broad general results are not affected by these large corrections. They emphasise, however, if that were necessary, that pioneering expeditions into unknown territory cannot be expected to bring back data as accurate as those obtained by a leisurely survey when the country has been opened up. It is now known that ionic charge is more nearly $1 \cdot 5 \times 10^{-19}$ coulombs.

We have seen that J.J. had proved that negative electrons could be got out of metals under the influence of ultra-violet light and out of an incandescent carbon filament. It was not a very far-fetched inference that they are given off by red-hot metals generally, though, owing to the complications introduced by charged atoms of metal or of gas, and the difficulty of using a high enough temperature, it was easier to assume this than to prove it.

In the meantime light came from another quarter. Several physicists, including J.J. Thomson himself, also Riecke and Drude in Germany, had developed the theory that conduction of electricity in metals was due to the presence in them of an atmosphere of mobile electrons, which was treated according to the methods of the kinetic theory of gases. The effect of the electric field on these was to superpose upon their random motions a definite average drift in the direction of the applied gradient of potential, thereby carrying the current. The energy of random motion of these free electrons will increase with temperature and, if the temperature is high enough, it will be enough to carry them through an obstruction at the boundary and out at the surface.

Thomson was the first to suggest in 1900 that this is in fact what does happen when a hot wire or carbon filament is found to give off electrons. However, he did not personally follow up the suggestion. It may be that his experience in the laboratory had taught him how difficult and perplexing a field of research this was likely to be.

O. W. Richardson* started work in the Cavendish Laboratory about this time and turned his attention to the question of emissions of electrons from hot wires, though not apparently getting any direct inspiration from the remark of Thomson above mentioned. He arrived independently at the same view, and developed it in quantitative detail, showing theoretically how to connect electron emission with temperature on the basis of these ideas. He showed that the theory would account well for his experimental results over a wide range and found that the work done in driving out an electron from platinum into vacuum was about equivalent to that required to move it over an opposing electromotive force of 4·1 volts. This work may be considered as putting on a firm basis the theory of electron emission in a modern thermionic valve. The word thermionic was coined by Richardson to cover phenomena of this class.

* Now Sir Owen Richardson, F.R.S., Yarrow Research Professor of the Royal Society.

CHAPTER VII

IN THE EARLY DAYS OF THE
TWENTIETH CENTURY

THE end of the year 1900 roughly marks the close of an epoch. Rutherford, when he came to Cambridge, already engaged to be married, had been heard expressing to J.J. some anxiety as to whether he would be able to get a suitable appointment when his scholarship expired, but J.J. had reassured him with the remark, 'Oh, there is always plenty of room at the top', and in fact he was appointed to the first professorship that became vacant—that at McGill University, Montreal. He was elected a fellow of the Royal Society shortly afterwards, but not so soon as J.J. considered he should have been, and J.J. betrayed his annoyance at this, criticising the composition of the council with whom the election practically rested.

Townsend, warmly recommended by J.J. and by Kelvin, went as professor to Oxford. Henderson had gone to become scientific secretary to Lord Kelvin, under whom he had been originally trained. Zeleny returned to America; McClelland, Langevin and Henry had also left. Others, notably C.T.R. Wilson and H.A. Wilson, remained, and new men, such as O.W. Richardson and (later) F. Horton, were coming on, so that there was no breach of continuity in developing the special line of research for which the Cavendish had become famous: nevertheless, the first wave of activity which followed the discovery of the X-rays had to some extent subsided, and this is a convenient point at which to break off from the narrative of research work, and speak of J.J.'s activities and interests in other directions. Efforts were made about this time to tempt him away from Cambridge to the Royal College of Science and to Columbia University, New York; but he does not seem to have seriously considered these offers.

In 1903 Thomson brought out his treatise on conduction of electricity in gases. It is most interesting to compare the account of the subject in this book with his account ten years earlier in *Recent Researches in Electricity and Magnetism.* It would be an exaggeration to say that the two accounts had nothing in common, for there is an account of the more obvious phenomena of the discharge in both: but by far the greater part of the new book was devoted to discoveries made in the meantime. Definite quantitative views could now be taken about the properties of the ions which conveyed the current in various circumstances, and the book remains a monument of the progress made in the short interval from 1896 to 1903. It may be said to summarise the work of Thomson's great days at the Cavendish Laboratory. A second edition came out in 1906, in the main reproducing the first edition. A final and greatly expanded edition of the book appeared much later in two volumes, published under the joint authorship of J.J. Thomson and his son, G.P. Thomson. J.J. explains in the preface that the work of preparing it had mainly fallen on the younger author. Nevertheless, the essential plan and a large part of the text of the original book are preserved, while the subject is brought up to date. The first volume of this edition appeared in 1928 and the second in 1933.

During the years 1900–1901, there was considerable fluttering in the dovecotes of the world of physics owing to the experiments of V. Crémieu, who had unsuccessfully attempted to repeat the experiment originally made by Rowland in 1876, by which he obtained a magnetic effect from electrical convection: that is to say, he had found that when electricity was carried round a circle by spinning an electrified disc, the magnetic effect was the same as that of an electric current in a flat wire coil coaxial with the disc, and in the same plane. (This is a rough general description only, enough for our purpose.) This matter came very near home for the workers in the Cavendish Laboratory: for if a moving electric charge did not produce a magnetic force, it would follow by the principle of action and reaction that a moving charge, such as the cathode stream was held to be, could not experience force in a magnetic field: and then the whole philosophy of the Cavendish School would have had its foundation shaken. Crémieu's experiments appeared so far as could easily be judged to have been well

designed and carefully carried out, and this made it, in the judgement of many of us, difficult to dismiss them summarily. J.J. never
seemed at all perturbed, and took the matter very lightly. G.F.
FitzGerald, whose opinion as a critic was generally held in high
regard, was inclined to think the position serious, but when J.J.
was confronted with his opinion he dismissed it as an example of
Irish perversity. Rowland had deduced an approximately correct
value of 'v' (the ratio of the electrical units) from his experiment.
'You cannot get a value of "v" out of a shake of the apparatus',
said J.J. This account of the matter was scarcely final, for Rowland
had never got his magnetic needle very steady and the deflections
were small. However, Rowland's conclusion was confirmed by
Pender, working at Baltimore, and a subsequent collaboration
between Crémieu and Pender cleared up the discrepancy, which
proved to be due to secondary causes.

As regards the class teaching in the laboratory, Thomson did
not himself take much part in the organisation of it. The course
had been arranged by Glazebrook and Shaw, working under Lord
Rayleigh, and since they remained in charge after Thomson succeeded him, matters went on as before. After they had left, others
took up the burden. Thomson of course made the appointments,
but when he had done so, he left his demonstrators to carry on
as they saw fit. No doubt he was quite sufficiently informed of
what was going on, and did not consider it necessary to interfere
in what was being well done by others. But I have heard him
imply, in rather a detached way, that the experiments in the advanced course were too hard, and that the pupils who could succeed in them would be better engaged in research. At the time
when I attended the classes in 1895–97, he used to come round
and speak to the students. I remember him finding me rather
muddled over some electrical connections, and he told me it was
his practice to draw them out on paper before beginning—a
valuable hint by which I have often profited. Later on, with the
increasing demands made on his time by the large number of
research students, he ceased to come round, and some of the staff
were inclined to complain that he neglected this side of the
laboratory's activities, and starved it financially. On one occasion

one of them bluntly attacked him for this, but he took this plain speaking with good temper, and got in another reluctant member of the staff to mediate. It is not every head of a department who would submit himself to criticism in this way.

His good-tempered and conciliatory attitude endeared him to the teaching staff. I have personally never seen him out of temper, or dangerous to approach, nor has anyone else told me of him in such a mood. He was the most accessible person in the world. He never seemed to be in a hurry, and was usually quite ready and even anxious to stop and discuss anything of public or scientific interest if one met him in the passage.

However, to return to the teaching staff. Those who served as demonstrators were, in order of appointment, H.F. Newall, later Professor of Astrophysics at Cambridge, of whom something has already been said; L.R. Wilberforce, an admirable teacher who became professor at Liverpool University in succession to Lodge; G.F.C. Searle, who devoted his life to developing the teaching of experimental physics at the Cavendish, and who (in spite of a temporary interlude of retirement) is still at it, at the time of writing, active as ever, and with undiminished enthusiasm is contriving new and ingenious experimental exercises after fifty-three years—a truly wonderful record.* He is one of the few who now remember Maxwell in the laboratory, though only on a visit during his boyhood.

Others were Sidney Skinner, who afterwards became head of the Chelsea Polytechnic: and T.C. Fitzpatrick, who undertook the entire management of the classes in physics for medical students. During part of the time he was President of Queens' College, and he only retired in 1916 when he became Vice-Chancellor.

Besides these, men took part in the teaching at various times whose names are better known in connection with research work, such as Threlfall, Callendar, Whetham,† Townsend, C.T.R. Wilson, G.I. Taylor. These names by no means exhaust the list.

* Perhaps I may be allowed to record my personal gratitude for the admirable teaching I received from Searle, Wilberforce and Skinner.
† Afterwards Sir William Dampier, F.R.S.

In his prime J.J. was very active of body; for example, if one was out for a country walk with him, he would lightly vault over a gate, while the rest of the party laboriously climbed it. On Sunday afternoons he might often be encountered walking by himself in the fields and lanes about Cambridge. On Saturdays he often arrived at the laboratory dressed for golf and in the afternoon took the train to Royston, where he played with any acquaintance who chanced to be there, or sometimes went round alone. On one occasion at least he took Rutherford, then recently arrived from New Zealand, with him and initiated him into the game. So far as I have been able to learn J.J. did not play a specially good game, but he was interested in the dynamical aspect of the subject, and once gave a lecture on it at the Royal Institution, when he used the transverse force on a cathode ray beam in a magnetic field to illustrate the dynamical effect of spin on the trajectory of a golf ball.

It is remembered by Cavendish men that J.J. was on one occasion in great form at tea. He said he was playing golf and drove against the wind. The ball went along mainly horizontally about a hundred yards and then rose up very high and came back over his head and fell behind him. An American there (H.A. Bumstead of Yale, it is believed) said very quietly, 'Is that a fact?' Everyone roared, including J.J.

I am indebted to Mr W. Craig Henderson, K.C., for a vivid glimpse of J.J. as a spectator on the football field at this period.

Townsend, McClelland, Henry and I were regular visitors to the Rugby football ground, when the University XV had a good home match (circa 1897), and there we were pretty certain to find J.J. and our interest was at times divided between watching the game and watching the Professor. He was an enthusiastic supporter of the game and has himself described a good passing movement, when the ball goes quickly from wing to wing and back again until the enemy's line is crossed, as 'the most thrilling thing in football or indeed in any game that I have ever seen'. Now when such a movement took place and the players dashed up the field, we would see J.J. sprinting along, outside the ropes, keeping up with the movement and determined to be 'up' when the try was scored. When on one such occasion he suddenly caught sight of us, smilingly watching him, he

checked himself in his stride, puckered his eyebrows in that little characteristic frown as if momentarily ashamed of being caught acting like
a young schoolboy, but was quickly off again. This again was but
another illustration of J.J.'s broad sympathies. There have been some
great scientists whose whole life was centred in their scientific work
and who had no interests outside it, but J.J. was not one of these, and
I feel sure that it was through his wide interests in all phases of human
activity that he was able to keep his brain so clear for his own exacting
labours so late in life.

J.J.'s description of the differences between football as played in
America and as played in England recalls to me the occasion when we
took John Zeleny to see his first football match played under the Rugby
code. After watching the game for some time he turned to me with
this remark: 'But—I—don't—smell blood!'

J.J.'s mind often worked in abnormal ways, with results which
caused a good deal of surprise. He sometimes seemed at the mercy
of an idea he had at the time, and then got quite out of touch with
his previous ideas and experience. It is conceivable that this
peculiarity was inherited from his mother (see p. 2). Several
instances could be given from different sources. Thus, for example, when Zeleny was investigating the velocity of gaseous
ions by methods which involved an air blast, J.J. remarked that
it would be of considerable interest to get similar data for metallic
vapours, mercury, zinc and so on. Zeleny assented, but so far
as he considered the matter seriously no doubt foresaw a campaign lasting over months, with much scheming to overcome
formidable difficulties of electrical insulation at high temperatures,
and many failures. He was therefore taken aback when J.J. said
to him a few hours later, 'Well, have you tried any metallic
vapours yet?'

Again, though I never saw anything of the kind myself, he has
been known to tell the same story twice over to the same company
at one sitting, which startled his hearers very much. This was in
his prime and had no sort of connection with any failure of his
powers. It must have been pure absence of mind.

Another instance of this, though rather out of its proper chrono-
logical place, may be mentioned here. At the beginning of 1919,
400 naval officers were to come to Cambridge. They had been taken

from Dartmouth very young for war service and after the war it was felt that something should be done for their education. They were to have lectures and Searle was to lecture on optics. J.J. came to his room one day with a list of lectures proposed for the officers, but Searle pointed out that it was unworkable, requiring him to give the same lecture three times over in three successive hours, and for the moment there was a deadlock. When Searle with other lecturers met the naval authorities a little later, and reported the plan proposed by J.J., he learned that that plan had already been abandoned by them on J.J.'s request, before J.J. had proposed it to Searle!

In 1899 the Thomsons moved from 6 Scroope Terrace to Holmleigh, West Road, on the Backs of the Colleges, where he remained until he became Master of Trinity in 1918. The lease of 6 Scroope Terrace was up, and Holmleigh had a much larger and better garden, which was the main attraction. Next to his family and the Cavendish Laboratory, the garden was very near J.J.'s heart. More will be said about this taste in a later chapter. Some building was done at Holmleigh a little later, which increased the size of the rooms.

The Thomsons were most hospitable. When they first married they gave breakfast parties for junior members of the University, but this had been given up when I came into residence in 1894. There were then tea parties on the days when Mrs Thomson was 'At Home' and dinner parties, including a varied company from the Master of Trinity (e.g.) down to a freshman like myself. Later, at Holmleigh, there were often dinners largely made up from the research workers at the Cavendish Laboratory. These included professors from American and other foreign universities some of whom were married and had their wives with them. There were, for example, at various times Lyman from Harvard, Bumstead from Yale, Duane from Colorado, Schmoluchowski from Lemberg, and many others.

It was whispered that J.J. did not always fall in with the necessity of those invitations which were dictated more by duty than by pleasure. I have said 'whispered', but it might be more accurate to say that it was shouted; at least that was the impression

J. J. Thomson in his study at Holmleigh, 1899. He is
sitting in a chair which was used by Clerk-Maxwell

of workers within earshot of the laboratory telephone, when J.J. was discussing the composition of a dinner party with Mrs Thomson. He was amusingly oblivious of the fact that he was furnishing all the dinner parties of Cambridge with enough scandal to last them a year!

The Thomsons from time to time entertained distinguished guests from outside Cambridge at Scroope Terrace or at Holmleigh. A list of some of these with dates may fittingly be given here. It does not aim at completeness, and it has not been thought necessary to distinguish those who stayed from those who came for lunch or dinner only.

1890   Sir William and Lady Thomson.
1891   Prof. and Mrs Ayrton; Mr Vernon Boys; Prof. A.W. Rücker, Miss Mary Kingsley; Prof. A. Cornu; Prof. H. Hertz.
1892   Prof. E. Wiedemann.
1895   Prof. G.F. FitzGerald.
1896   Prof. J.S. Ames; Prof. and Mrs A. Schuster.
1897   Miss Newton (fiancée of E. Rutherford).
1898   Prof. and Mrs Jean Perrin.
1899   (For the Jubilee of Sir G.G. Stokes as Lucasian Professor): Prof. A. Cornu; Prof. R.B. Clifton; Prof. S.P. Langley; Prof. Oliver Lodge; President and Mrs Patton of Princeton; Sir Andrew Noble.
1902   Prof. and Mme Henri Becquerel.
1903   Prof. Simon Newcomb; Prof. P. Curie.
1904   (British Association Meeting in Cambridge): Dr J. Elster; Dr H. Geitel; Prof. and Mme Langevin; Prof. and Mrs J.H. Poynting.
1905   Prof. and Mrs W. Duane; Prof. H.N. Russell.
1906   Prof. and Mrs Zeeman; Prof. and Miss Runge.
1907   Prof. and Mrs G.E. Hale.
1908   (On the occasion of the Installation of Lord Rayleigh as Chancellor of the University): Sir William Crookes; Sir Andrew Noble; Prof. J.H. Poynting.
        And at other times Prof. S. Arrhenius; Prof. and Mrs H.A. Lorentz.
1915   Duc de Broglie.

J.J., I am inclined to think, had a very good nose for any kind of swindle or hoax, and this may have some connection with his success as an investor on the Stock Exchange. The Cavendish was one day visited by a platinum thief, posing as an American professor. He knew enough not to be detected immediately by his technical ignorance; but J.J. decided at once that he was an impostor on the ground that 'that was not the kind of person an American professor was', and he did not make any haul.

Again, a group of post-graduate students, including Rutherford, Townsend, McClelland, H.A. Wilson and some others, tried experiments on thought transference. Townsend and McClelland apparently had the power to do it, and successfully hoaxed the others. Someone wrote a paper for the Psychical Research Society, and J.J. was told about it. He grinned, and said, 'Oh, yes, two Irishmen'. They confessed shortly afterwards.

We have not said much about the development of radioactivity. J.J. was keenly interested in the subject, though he did not at any time take a conspicuous part in the development of it. English experimenters were at a disadvantage in this matter, because during the early period, when the field was virgin, there were no such highly concentrated specimens of radium in this country as there were in France, Germany and Austria. Rutherford's early work in 1899 was carried out in the Cavendish Laboratory. Using uranium and thorium as his sources, he showed that the ions produced in air had the same properties as the ions produced by X-rays. He further distinguished between the α- and β-rays and gave these initials to the rays in question, by which they have been designated ever since. After Rutherford left for Montreal, J.J. was in regular correspondence with him. I remember him coming in one day full of enthusiasm about the thorium emanation of which he had just heard by letter. He poured out the whole story to Townsend and myself and as soon as he had gone Townsend said (truly enough): 'He is like a child with a new toy with his emanation.'

E. RUTHERFORD *to* J.J. THOMSON *from Montreal. Jan. 9th*, 1900:

The results of Giesel and Becquerel (on the magnetic deflection of β-rays from radium) are very interesting and remarkable. I expect the 'emanation' in thorium is also true for polonium and radium when prepared in a special manner, and that the deflection due to the magnetic field is due to the action on a charged particle cast off from the active body.

This letter recalls how strange the deflection of the β-rays of radium by a magnetic field appeared when first observed. The straightforward interpretation of it now adopted did not at first seem possible to Rutherford, nor as I clearly remember to J.J. Thomson either. Lenard had got a stream of electrons through thin metal foil, but it still seemed incredible that they could get through metal half a millimetre or so in thickness.

Henri Becquerel, the discoverer of radioactivity, visited Cambridge in March 1902, staying with the Thomsons at Holmleigh. They invited Sir George Stokes and others to meet him. He visited the Cavendish and showed a powerful specimen of radium, which was an exciting novelty to the workers there, who had only had access to weak preparations. Becquerel, it was said, developed a serious lesion on his stomach beneath the waistcoat pocket in which he was accustomed to carry his specimen of radium about, the danger of it being then unrecognised.

J.J. showed him round the Cavendish Laboratory. Becquerel talked French. J.J. as usual made no attempt to reply constructively in that language, and said 'Oui, oui' in a loud voice after everything Becquerel said. It could be heard all over the laboratory (not then so extensive as now).

The sequel is amusing. Sometime later an Indian physicist came and J.J. showed him round. He said 'Yes, yes' after everything J.J. said. When they got to the room where H.A. Wilson was working, J.J. introduced him to the Hindoo, and asked him to show him the rest. Wilson equally failed to elicit any other comment. Next day J.J. told Wilson that he could not stand any more of the man's 'Yes, yes'.

In May 1903 J.J. again visited America to give the Silliman Lectures at Yale University. He travelled alone, Mrs Thomson being unable to go. His impressions of Yale are given fully in his *Recollections*.

From the Graduates Club, New Haven, Connecticut. May 22nd 1903.
To G. P. Thomson:

The son of one of the Professors here who is only seven years old wanted to attend my lectures, but was told that he would not understand them. I met him and had a talk with him one afternoon. When I had gone he told his mother he thought it was a great shame he had not been allowed to go to the lectures for he had had a talk with Professor Thomson and could understand what he said as well as he could anyone else.

Afterwards, in June, he visited his great friends, Prof. and Mrs H. F. Reid, at their summer house at Monte Vista, Blue Ridge Summit, near Gettysburg, Pennsylvania. He returned to Cambridge towards the end of June.

The Silliman lectures were published with very little delay in a volume entitled *Electricity and Matter*. Like the previous volume giving his Princeton Lectures, this is now mainly of historical interest and only a few of the topics can be noticed here.

In the early chapters he deals with Faraday's tubes of force, and attempts to develop this conception much beyond the use generally made of it by Faraday and Maxwell in describing electric and magnetic attractions and the induction of currents. He introduces this conception into atomic science, and pictures the emission of light waves from individual atomic centres as spreading along discrete Faraday tubes, which are no longer used statistically, but are taken to have real physical individuality, like stretched strings. He had at this time definitely adopted the theory that Röntgen rays were light waves of short wave-length, and he discusses how the sudden stoppage of cathode-ray corpuscles would send out a sharp kink—like a pulse along the tube of force attached to the corpuscle. This pulse was considered to constitute the X-ray. Conceptions of this kind are not now in date, because they meet with those difficulties which all electromagnetic theories meet with, when it is attempted to apply them to radiation phenomena on an atomic scale—difficulties which have compelled the adoption of the quantum theory.

Thomson also suggests that this theory of tubes of force eases a serious difficulty in the field of X-ray ionisation which had not before been emphasised. When the X-rays fall on a gas, they ionise some of the molecules. But the ionised molecules are only an infinitesimal proportion of the whole number, less than one in a thousand million, and the question presses—what it is that causes these particular molecules to be selected and the others not touched? He gives cogent reasons for thinking it is not any peculiarity in these particular molecules in respect of the amount of energy they possess, and then goes on to suggest that the Faraday tube theory of radiation will afford an explanation, since it indicates a wave front which is not uniform, but has specks of great intensity which are able to ionise.

There is, however, a fundamental objection to this theory which oddly enough does not seem to have troubled J.J. at this time, though he made an attempt to deal with it several years later. It must here be briefly explained.

What reason have we for thinking that light consists of waves at all? The answer is that the hypothesis explains in close detail the phenomena of interference and diffraction. A typical example is Young's interference experiment. In this experiment light from a pinhole source passes through two close pinholes or slits in a plate placed some inches distant from the first pinhole and then falls on a screen. In the symmetrical position there is a bright band. Passing to either side, there is a dark band, and then a bright band again, and so on for several alternations. Where a dark band occurs there is destructive interference of light, that is, the light from one of the pinholes destroys the effect of the light from the other pinhole: and this may be verified as Young did by covering one of them up when brightness returns. This experiment receives its explanation on the wave theory by the principle of interference, the crest of one wave falling into the trough of the other, and thus neutralising it. But if the wave front is speckled or spotted according to J.J.'s view, what becomes of this explanation? Unless the spots in both coincide on the screen, there is no scope for interference to take place; and why should they coincide, when they come from different directions? If they are numerous enough to overlap one another, then the advantage for the sake of which the theory was proposed, is lost. We can therefore only obtain the spotted wave front by rejecting the main reasons for believing that a wave front exists at all. To what extent J.J. may have been troubled by this in 1904 is not very clear. He was very prolific in explanations of the difficulty which immediately confronted him, but it was not perhaps in accordance with his habit of mind to work round the horizon to see what other difficulties he might be introducing. There was nothing of the defeatist in his make-up, and he usually preferred to dwell on what a theory *would* explain than on what it would not. Other thinkers have been reluctantly forced into a somewhat similar position, but they have admitted the dilemma more explicitly than he did. He did make an attempt later to repair the theory in this respect, but with little success.

There is in fact no satisfactory explanation of the difficulty which troubled J.J. about why one molecule is selected to be ionised when millions of others are not so selected. On the corpuscular theory of light which was held in the eighteenth century no particular difficulty arises. On this theory a molecule is ionised if it is lit by a light corpuscle, otherwise it is not ionised. But the other phenomena of interference which early in the nineteenth century defeated the corpuscular theory of light remain unexplained.

The best synthesis which has been made of the conception of a uniformly spreading wave with the contradictory fact that its effects are highly localised at certain places in the wave front is found in the conception of a *wave of probability*. Some philosophers object to this phrase, which they consider a piece of meaningless mystification, but after all it is only intended to summarise observed facts.

To begin with, what is meant by a wave? It means a *state of things which is transmitted*. In the case of a sound wave, there is no difficulty in specifying what this state of things is. It is a local compression of the air. In the case of light, it is much more difficult to specify the nature of the state of things transmitted.

To get some idea of a wave of probability consider a reckless motorist moving along a road. There is a certain probability of an accident where he is. If he moves on, the probability of an accident moves also. Thus the moving probability satisfies the definition of a state of things which is transmitted and may properly be called a wave of probability. The reader may compare this with the bow wave of a steam launch, which accompanies the launch just as the probability wave of an accident accompanies the reckless motorist. We can have a water wave without the launch, and similarly we may imagine a wave of probability without the motorist. When such a wave comes up to a molecule there is the (remote) probability of a catastrophe, which takes the form of the molecule being ionised. The more intense the wave, the greater this probability is, but the intensity of the wave does not affect the extent of the catastrophe, but only the probability of the standard catastrophe occurring. So far as I know this conception of a wave of probability is not put forward by anyone as satisfying. It is merely descriptive of what is observed to happen. It is as if nature were a conjuror. We describe her trick as it appears to us, without professing to see through it. J.J. was then one of the first, perhaps actually the first, to emphasise the particular difficulty about the localised action of a wave on matter. It turns up in various forms in modern physics, and is more commonly expounded in connection with the photoelectric effect, i.e. the liberation of electrons from metals. He did not succeed in proposing a satisfying solution, and it may be questioned whether anyone else has been more successful. Any strictly deterministic treatment of the subject seems to be impracticable.

In the course of these lectures Thomson attempted to develop further the probable structure of the atom, on the basis of the ascertained fact that it contained corpuscles or electrons. There was not much to go on beyond the bare fact that all kinds of atoms apparently contained the same kind of electrons, and the theory which Thomson built up from this fact was no doubt very speculative. Though it cannot be considered to stand in the condition of science to-day, yet it contained much which contributed to and foreshadowed our present notions.

To begin with, the general conception that the atom has a certain resemblance to a planetary system is used. I do not know who first suggested this comparison, but the facts of spectroscopy were enough to prove that the structure of the atom could not be simple, and the comparison is probably at least some decades earlier than Thomson's time, but it hardly amounted to more than a phrase. Prout had supposed that the constituent members of the more complex atoms were atoms of hydrogen. Thomson considered them to be electrons. We now think that there was truth in both notions. However, we are considering Thomson's contribution.

If the atom was a complex system containing electrons, these latter might be supposed to be describing orbits, or to be held stationary. The latter hypothesis was preferred, not so much on the ground that it was more likely as that it seemed more manageable. If the electrons were stationary in the atom, the structure must be such that it would give them stable equilibrium: that is to say, if an electron were displaced from its position, there must be a force tending to restore it to that position. The electrons being negatively electrified, they could only be held by positive electrification, and the question arises how this positive electrification can be supposed to be distributed so as to keep the electrons in stable equilibrium. This is a form of what is sometimes called the problem of 'Mahomet's coffin', which according to legend is supposed to float between heaven and earth without touching anything. This result cannot be got out of forces which vary as the inverse square of the distance. We cannot hold an electron in equilibrium by any forces arising from electrified bodies at a distance from it—as may readily be proved from the theory of attractions. On the other hand, it can be done if we put the electron inside a uniform distribution of electrification.

Thomson made use of this conception, which, as he mentioned, had been used by Lord Kelvin a little earlier. He supposed a sphere with positive electrification uniformly distributed inside it, and he placed his electrons inside this sphere, leaving them to find their positions of equilibrium under their mutual repulsions and the attraction of the positive electricity.

If there is only one corpuscle it will place itself at the centre of the sphere. If there are two, they will be in equilibrium at equal distances from the centre, along a diameter. If there are three, an equilateral triangle meets the case, if four a regular tetrahedron.

In these cases the corpuscles rest in equilibrium, and they are all at the same distance from the centre, lying as we may say on a single shell. Thomson was able to show that when the number was greater, say seven or eight, this could no longer be the case. The corpuscles distribute themselves over two concentric shells; and with a certain further increase, three shells become necessary.

In these more complicated cases the theoretical problem becomes unmanageable, and Thomson appealed to certain experiments made with magnets by Prof. A.M. Mayer of the Stevens Institute of Technology, U.S.A. In these experiments (originally made about 1878) a long bar magnet was held vertically over a bowl of water, and on the water a number of thin permanent magnets made from needles floated in corks. The magnets were long enough for only those poles which were near to the water surface to count. The upper pole of the fixed bar magnet, and the under poles of the floating magnets, were thought of as far enough off not to be of much account. The acting pole of the fixed magnet was positive, and of the floating magnets negative,* and these poles then became the analogues of the positive charge and its surrounding electrons. The constraint introduced by the flotation secures stable equilibrium without the device of a sphere of positive attracting matter in which the electrons are placed.

The important feature of these experiments is that they show the formation of successive rings of magnets. For example, any number of magnets up to five will arrange themselves at the angles of a regular polygon, but beyond this number they will not do so, and a new inner group begins to form itself by one magnet occupying a central position. This group develops as magnets are added until with fourteen magnets there are nine outside and five inside; when a third group begins to develop, this is complete when there are twenty-six magnets.

---

* They might of course be reversed. I have adopted the above convention for obvious descriptive reasons, but there is no distinction between positive and negative magnetism analogous to the distinction between positive and negative electricity.

J.J. emphasised these experiments as probably giving the key to the periodic law in chemistry, which, broadly speaking, states that the properties of the chemical elements are a periodic function of the atomic weights, just as the structure of the magnet pattern is a periodic function of the number of magnets thrown in. He used to show these experiments in his elementary lectures some years earlier and explain his ideas about them in relation to the periodic law. I think he did this before he had the electron idea at all. It was rather too strong meat for some of his students and I remember a fellow-student remarking to me that he thought it altogether fanciful.

The modern view derived by the detailed study of spectra on the principles developed by Bohr and his school is very like this, and far more definite. It affirms very definitely, for example, that the rare gases correspond to completed rings of electrons. Take neon as an example. The preceding electro negative element fluorine contains one electron short of the complete ring, and the electropositive sodium which follows neon has one more, which cannot find a place in the ring. It would be tempting to read all this into J.J.'s account, and he seems to be rather near it; but not quite there. He was no doubt at a disadvantage in that the periodic law really refers to atomic numbers. It was in those days formulated for atomic weights, which do not follow quite the same order, and cannot be forced into the scheme without some 'cooking'. His ideas can only be considered suggestive. The magnets were only a rough model of his model of the atom: and the atom itself was doubtless far from either. Nevertheless he did open up a new train of thought: and it has in a broad sense been singularly justified by events.

To Sir Oliver Lodge: *April 11th*, 1904.

With regard to positive electrification I have been in the habit of using the crude analogy of a liquid with a certain amount of cohesion, enough to keep it from flying to bits under its own repulsion. I have, however, always tried to keep the physical conception of the positive electricity in the background because I have always had hopes (not yet realised) of being able to do without positive electrification as a separate entity, and to replace it by some property of the corpuscles. When one considers that all the positive electricity does, on the corpuscular theory, is to provide an attractive force to keep the corpuscles together, while all the observable properties of the atom are determined by the corpuscles, one feels, I think, that the positive electrification will ultimately prove superfluous and it will be possible to get the effects we now attribute to it, from some property of the corpuscles.

At present I am not able to do this, and I use the analogy of the liquid as a way of picturing the missing forces which is easily conceived and lends itself readily to analysis.

As will be gathered from the above letter J.J. was not inclined to be dogmatic about his atomic theories, and indeed he was quite prepared to change them, sometimes without making it altogether clear that he had wiped the slate clean, and that what he had written before must now be considered cancelled. One correspondent (Mr E. Pickworth Farrow) has told how much struck he was with J.J.'s answer to some objection: 'Well, of course I have always thought that the chief value of any theory was as a basis for further experiments.'

The following interesting letter from Lord Kelvin foreshadows the insuperable difficulty that has been found in a determinist theory of the breaking up of radioactive atoms:

15 Eaton Place, S.W.   12th Nov. 1906.

Dear Thomson,

Thanks for your letter of yesterday. It seems to me that the supposed store of fifteen hundred years energy in a radium atom, in the form of kinetic energy of electrons, or of 'corpuscles', both vitreous and resinous, moving within it, would require velocities far exceeding the velocity of light.

It seems to me a fatal objection to that theory that it implies each radium atom, with the 'vast number of corpuscles' moving in its interior, should be predestined to come to the instability which causes them to be shot out at a predetermined time. When did the history of this state of affairs begin for each atom? and what external influence gave it its moving corpuscles? What would be the difference, between radium atoms in a piece of radium bromide, of the performance of those of the atoms which are nearly ripe for explosion, and those which have the prospect of several thousand years of stable diminishing motions before explosion?

The whole subject put before us by the astonishing results of experiment is too frightfully difficult. I have no doubt however but that in a few years much more will be known about it, and it will be even more interesting than ever.

Yours truly, KELVIN.

It would be interesting if we had J.J.'s answer to this letter, but in default of it a passage may be quoted from his British Association address at Winnipeg three years later.

The evidence we have at present is against a disturbance coming from outside breaking up the radioactive atoms and we must therefore look to some process of decay in the atom itself, but if this is the case, how are we to reconcile it with the fact that the expectation of life of an atom does not diminish as the atom gets older? We can do this if we suppose that the atoms when they are first produced have not all the same strength of constitution, that some are more robust than others, perhaps because they contain more intrinsic energy to begin with and therefore have a longer life. Now if when the atoms are first produced there are some which will live for one year, some for ten, some for a thousand and so on; and if lives of all durations from nothing to infinity are present in such proportions that the number of atoms which will live longer than a certain number of years decrease in a constant proportion for each additional year of life, we can easily prove that the expectation of life of an atom will be the same whatever its age may be. On this view the different atoms of a radioactive substance are not in every respect identical.

This passage is interesting as an illustration of J.J.'s mental reaction against a difficulty which most people have considered insuperable. He was so fertile in suggesting ways of getting out of a conclusion he did not want to accept, that the idea of a crucial test of anything was apt to fade away in talking matters over with him. No doubt the hypothesis he here introduces will serve the purpose for which it was invented, but it is so special as to make it doubtful whether it can have satisfied its author or anyone else, and in fact it is almost evident from his words that he was not satisfied with it.

The modern view is that there is no possibility of making a distinction between those atoms which are ripe for explosion and those which have the prospect of several thousand years of life before explosion. It is easier to deal with radon, the atoms of which do not last nearly so long, but the principle is the same. If we start with a portion of radon—a limited population of radon atoms—we shall find that half of them have exploded in four days. Have the remainder been effectively growing old in the meantime—have they made progress towards the inevitable explosion? Not at all. If they had their expectation of life would clearly be diminished, but when we make the test we find that only half of *them* are exploded in four days. Compare this with a human population, which, from the moment of birth, is immediately making progress towards the inevitable end. Half of it, say, is dead in fifty-five years after birth—I do not pretend to be precise, not having exact data at hand. What then do we find after another fifty-five years? Are half of the first batch of survivors, or a quarter of the original number, still surviving? No, they are all dead. We see that the radioactive atoms are quite different. Their explosion is a fortuitous event, and not the result of causes working up to an inevitable end.

J.J. in his *Conduction of Electricity through Gases* considered the problem of the scattering of X-rays by electrified particles. The X-ray is considered as a thin pulse of electromagnetic radiation, and the particle is accelerated under the electric force in this pulse, thereby becoming a source of scattered radiation, issuing in a direction lateral to the primary beam. This calculation assumes the particles to be free, and was originally made with a view to the scattering of X-rays by ions. It shows how much energy the scattered radiation ought to have relatively to the original radiation, supposing the mass, charge and population density of the scattering particles to be known.

In the meantime C. G. Barkla, originally a pupil of Sir Oliver Lodge, had been working, at Thomson's suggestion, on the secondary radiation from gases exposed to X-rays. This work was begun at Cambridge, and afterwards developed independently at Liverpool and elsewhere. Thomson, however, continued in touch with it, and Barkla has put on record how great a stimulus was given by him. Barkla found that the secondary radiation from 1 c.c. of air was 0·00024 of the primary radiation. Applying Thomson's theory of scattering to this result, he found it impossible to explain it by the action of ions associated with atomic mass, because particles so heavy could not move fast enough to give the observed amount of scattered radiation. If the attempt were made to compensate for their sluggishness by their great number, then we should have to postulate enormously more ions than there were atoms or molecules present. Barkla argued therefore that the scattered radiation could not be produced by atomic ions, and he fell back on the supposition that it was produced by electrons in the atom, which he assumed were free enough to act as independent scattering agents. Putting in the then accepted values for the mass and charge of an atom, he got a result indicating about 100 per atom, or say four times the atomic weight.

The result is, however, very dependent on the accuracy of the atomic constants used in the calculations, any error in these telling several times over: and the errors in the early values were large. This was the only cause of Barkla's comparatively unsuccessful result. J. J. Thomson, using the same scattering data, and better values for the atomic charge and mass, concluded that the number of scattering electrons was equal to the atomic weight: but the modern values for the atomic charge and mass would give, on Barkla's data, more nearly half the atomic weight, and this is in complete agreement with modern views. Thomson emphasised strongly that the number of electrons in the atom was of the same order of magnitude as the atomic weight, and this conclusion rests on his calculation of scattering, combined with the experimental work of Barkla, and the recognition by the latter that this method could give the number of electrons in the atom. Since it would take about fourteen thousand electrons to make up the mass of a nitrogen atom, it became clear that the moderate number of electrons present—whether seven or fourteen—could not go far to make up the mass, and the earlier idea that the atom was composed of electrons was explicitly given up by Thomson at this stage. The existence of the atomic nucleus, small and massive, was not recognised until considerably later. It came out of the investigations of Rutherford and his school on the scattering of α-particles from radioactive bodies.

In 1902 J.J.'s pupils subscribed for the portrait by Mr Arthur Hacker which hangs on the stairs of the Cavendish Laboratory. They were delighted to see that the characteristic neglect of the details of his toilet was faithfully reproduced: and apparently he did not mind himself, since this was said to be the portrait he liked best.

The Cavendish dinner, the inauguration of which has already been described (p. 64), became an established institution. Songs written in parody of well-known popular airs and dealing with the studies and personalities of the laboratory were written and sung on the various occasions. The majority of these were by the late A.A. Robb, F.R.S., of St John's, at one time a worker in the laboratory, and the author of a treatise on 'Time and Space'. He inherited landed property in Northern Ireland, but he continued to reside partly at Cambridge. One would hardly have suspected his poetic gifts. The following is to my taste one of his best efforts:

The Don of the Day
Air: 'Father O'Flynn.'

1. Of Dons we can offer a charming variety,
   All the big pots of the Royal Society,
   Still there is no one of more notoriety
   Than our professor, the pride of us all.
   Here's a health to Professor J.J.!
   May he hunt ions for many a day,
    And take observations,
    And work out equations,
    And find the relations
     Which forces obey.

Chorus: Here's a health to Professor J.J.!
   May he hunt ions for many a day,
    And take observations,
    And work out equations,
    And find the relations
     Which forces obey.

2. Our worthy professor is always devising
   Some scheme that is startlingly new and surprising,
   In order to settle some question arising
   On ions and why they behave as they do.

Thus, when he wants to conclusively show
Some travel quickly and some travel slow
 He brings into action
 Magnetic attraction
 And gets a deflection
  Above and below.

3. All preconceived notions he sets at defiance
By means of some neat and ingenious appliance,
By which he discovers a new law of science
Which no one had ever suspected before.
All the chemists went off into fits;
Some of them thought they were losing their wits,
 When quite without warning
 (Their theories scorning)
 The atom one morning
  He broke into bits.

4. When the professor has solved a new riddle,
Or found a fresh fact, he's fit as a fiddle,
He goes to the tea-room and sits in the middle
And jokes about everything under the sun.
Then if you try to look grave at his jest,
You'll burst off the buttons which fasten your vest,
 For when he starts chaffing,
 Your tea you be quaffing,
 You cannot help laughing
  Along with the rest.   A. A. R.

The song that follows was of composite authorship. H.A.
Wilson wrote several verses of it, and showed it to J.J. He took
it home and brought it back in the morning with a new verse;
the fourth. Whether all the rest was by Wilson I am not sure;
I have seen it credited to J.J.E. Durack. J.J. asked for a copy
to deposit in the laboratory archives, where presumably it still is.

Ions Mine

Air: 'Clementine.'

1. In the dusty lab'ratory,
  'Mid the coils and wax and twine,
There the atoms in their glory
  Ionise and recombine.

Chorus: Oh my darlings! Oh my darlings!
    Oh my darlings ions mine!
  You are lost and gone for ever
    When you just once recombine!

2. In a tube quite electrodeless,
    They discharge around a line,
  And the glow they leave behind them
    Is quite corking for a time.

3. And with quite a small expansion,
    $1·8$ or $1·9$,
  You can get a cloud delightful,
    Which explains both snow and rain.

4. In the weird magnetic circuit
    See how lovingly they twine,
  As each ion describes a spiral
    Round its own magnetic line.

5. Ultra-violet radiation
    From the arc or glowing lime,
  Soon discharges a conductor
    If it's charged with minus sign.

6. α rays from radium bromide
    Cause a zinc-blende screen to shine,
  Set it glowing, clearly showing
    Scintillations all the time.

7. Radium bromide emanation
    Rutherford did first divine,
  Turns to helium, then Sir William
    Got the spectrum—every line.

The songs were privately printed under the title 'Post Prandial Proceedings of the Cavendish Society' in 1904, and were afterwards published. There were in all six editions. The last was published in 1926.*

In 1905 J.J. was appointed Professor of Natural Philosophy at the Royal Institution, on the resignation of Lord Rayleigh. This post was one which had a high tradition. Thomas Young and John Tyndall had held it, and Tyndall's work had mainly been done

* A commentary upon them by Prof. John Satterly will be found in *The American Physics Teacher*, Vol. VII (1939), pp. 179–85 and 244–8.

in the laboratory attached to the professorship.* Rayleigh had used it for only part of each year, and that this would be so was clearly understood when he originally accepted the post. An important part of his work on argon had been done there. However, he did not consider that the laboratory was being adequately used, and this was one of his reasons for resigning. He felt therefore a certain sense of frustration when J.J., who obviously would not use the laboratory at all, was appointed to succeed him. From another point of view he no doubt considered the appointment an excellent one, for no one could better interest the Royal Institution audiences with afternoon and Friday evening experimental lectures than J.J. No doubt this obvious consideration weighed with the Managers. They were, however, a good deal under the influence of Dewar, the resident Professor of Chemistry; and it may be suspected that he preferred to have the limited resources of the laboratories to himself. In saying this it is not intended to imply that he did not make a good use of them.

J.J.'s appointment at the Royal Institution carried with it the services of an assistant, and it was agreed by the Managers that the assistant should be available not only at the Institution but wherever the Professor might want him. Thomson was glad of the opportunity of getting additional help, and he decided to appoint a man of academic qualifications, who would be complementary to Everett. Everett, though excellent on the purely mechanical side, had little understanding of science—and indeed, did not wish for any; his object was to do what the Professor wanted, and he did not aspire to understand what the Professor was aiming at, and even cut anyone short who attempted to explain it to him. The late Dr G. W. C. Kaye† was appointed. He came from the Royal College of Science, where he had been trained under Rücker. From that time on, J.J. gave at the Royal Institution an annual course of six afternoon lectures open to the public by payment, as well as a Friday evening to the members and their friends.

* Since the reconstruction of the building in 1929 this laboratory no longer exists in recognisable form.
† Afterwards Superintendent of the Physics department of the National Physical Laboratory, F.R.S., died 1941.

The experiments for J.J.'s lectures in London were for the most part prepared and rehearsed at Cambridge, and the apparatus, or at any rate such parts of it as were at all special, was taken up to town by Everett, who with Kaye and later Aston attended to help with the demonstrations. He tried to seek out anything of novelty for his London audience; he once or twice asked me to lend him slides or to give hints for experiments if I had been doing anything which would suit his purpose; and doubtless other of his friends and pupils had the same experience. The only record of his Saturday afternoon course which survives is in the excellent reports of some of them in *Engineering*, due to Mr H. Borns, a capable professional scientific reporter and journalist of that time. These reports were often looked over and if necessary corrected by J.J. himself, but he had no hand in drafting them.

The courses in successive years from 1905 to 1920 inclusive were: Electrical Properties of Radioactive Substances. Corpuscular Theory of Matter—Röntgen, Cathode and Positive Rays. Electric Discharge through Gases. Properties of Matter. Electric Waves and the Electro-magnetic Theory of Light. Radiant Energy and Matter. Molecular Physics. The Properties and Constitution of the Atom. Recent Discoveries in Physical Science. Recent Researches on Atoms and Ions. Radiation from Atoms and Ions. The Electrical Properties of Gases. Problems in Atomic Structure. Spectrum Analysis and its Application to Atomic Structure. Positive Rays.

It will be seen that these lectures were for the most part on the subjects on which he specialised, and no doubt contained much of the matter which he was accustomed to give at Cambridge, somewhat popularised to suit a more general audience. It was and is the custom of the Royal Institution that blackboard mathematical calculations are not to be introduced.

J.J.'s regard for the views of those members of his audience who carried special weight was sometimes evidenced in his Friday evening lectures at the Royal Institution. If Kelvin or Rayleigh were present he would, at crucial points, appear to be addressing his remarks to them. He would go over to their end of the lecture table and explain his point with gesticulations, and for a few moments would appear to be oblivious of the rest of his audience. To appear to address some remarks in a lecture to one or two members of a class was, however, not an uncommon habit of J.J.'s. It was at times a little disconcerting to the individual singled out and on one occasion a certain American professor felt called upon to say something by way of reply, a reaction which seemed to surprise J.J. as much as it did the class.

The number of those engaged in research grew very rapidly and for many years before the war there were always thirty or more researches going on in the laboratory. This put a great strain upon the apparatus and upon the workshop. It also necessarily absorbed a great deal of J.J.'s own time to interview so many almost daily, even though he did not seem to be exhausted intellectually by doing so.

The preparation for his lectures became more sketchy than formerly and poor Hayles, the lecture assistant, might be seen running after him trying to find out what experiments he wanted for his lecture, sometimes almost in despair. Then J.J. would blurt out 'Diffusion of Gases' and Hayles had to do the best he could on that indication.

Sometimes, when he was called away to London, he would ask Searle, Horton, or some other member of the laboratory staff to take his lecture for him, usually giving very short notice. It was suspected that he did this on purpose, to prevent too much time being spent in preparation.

J.J. liked his research men to propose their own problems, but it was only a minority who could make a start in that way and there were heavy demands on him for suggestions when so many were at work. He told me once that he had no difficulty in finding problems of a kind—the trouble was to find reasonably easy problems for beginners. He seldom discouraged anyone from trying anything, however wild the idea might seem, or however much he might judge it beyond the technical capacity of the man who wanted to do it. He probably thought it best to let them find that out for themselves, and after all, prevision on such matters is very apt to be wrong.

It was his boundless enthusiasm, his endless fertility in suggestion, and his unequalled knowledge of the literature that made him such an inspiring teacher. This last has not perhaps always been sufficiently appreciated. He would refer to some incidental remark concerned with something perfectly different, and published ten years ago, and would add: 'I think you will find it about the middle of the fourth page, opposite to such and such a diagram.' He also claimed to be able to recite the names of any past University Cricket or Rugger team: but I believe that those who took the trouble to check him sometimes found that he was over-confident on these matters. There were serious anachronisms in his description of the cricket matches of the past.

Another quality of J.J.'s which impressed those who were brought into contact with him was his quickness at mental arithmetic—not the infallible accuracy of the 'calculating boy' but the ability to carry out numerical applications of quite complicated algebraical processes in his head with sufficient accuracy. He maintained that a slide rule was a waste of time. When one of his pupils challenged him to a trial of his methods against the slide rule, he left his challenger standing at the post.

The converse of J.J.'s amazing fertility of suggestion was an occasional lack of receptivity of the ideas of others. This was certainly not his normal habit of mind, but in some moods it was definitely in evidence. When some idea was suggested to him, he would be inclined either to come back the next day and repeat to its author what he had been told the day before as if it was new, or he would not accept it at all. He seems, for example, to have acquired a fixed antipathy to the practice of 'out-gassing' surfaces of glass and metal in which it was desired to maintain a high vacuum, because he had once taken up this attitude.

His mind was working incessantly. If an idea struck him in the laboratory, he would bend forward a good deal, rub his hands together vigorously and dart across the room as far as the size of it would allow. Perhaps it was this impact of a new idea that would suddenly make him change his motion from a walk to a jump when he was going along King's Parade with his hands in the pockets of his veteran overcoat, sometimes chuckling to himself. We might perhaps laugh at his little peculiarities; but we knew he was a great man, and we all loved him.

As an example of his absence of mind: Everett had very carefully made several small thin metal discs of different metal but all of the same diameter and weight. J.J. was talking to him about the experiment for which the discs were required, and he picked them up one by one and rolled them into little balls. Then he put them down and walked away quite happily. Poor Everett began making a fresh set.

Sometimes when things were not going well a little tension was apparent in the relations between J.J. and Everett; Everett being convinced that J.J.'s view of what should occur was wrong, and J.J. sure that the failure of the experiment was due to Everett not having carried out his instructions properly. On one occasion J.J. was attempting to deflect by means of a magnetic field positive rays produced in a glass globe some six inches in diameter, situated between the poles of a large Du Bois electromagnet. J.J. had his eye to a microscope and was observing the motion of the gold leaf of an electroscope across a scale. The conversation was overheard.

> *J.J.* 'Put the magnet on.' Then followed a click as Everett closed a large switch.
> *J.J.* 'Put the magnet on.'
> *Everett.* 'It is on.'
> *J.J.* (eye still to the microscope). 'No it isn't on. Put it on.'
> *Everett.* 'It is on.'

A moment later J.J. called for a compass needle. Everett went out of the room and returned with a large needle, 10 inches long, which was used in elementary lectures in magnetism. J.J. took it, and approached the electromagnet. When about a foot away, the needle was so strongly attracted by the electromagnet that it swung round and flew off its pivot, crashing into the bulb (which burst with a loud report) and coming to rest between the poles of the magnet. The spectators looked up to see what had happened. Everett was glowing with triumph, and J.J. looking at the wreck with an air of dejection. 'Hm,' he said. 'It was on.'

There are three eyewitnesses to this incident. The variations between them are trifling.

CHAPTER VIII

NOBEL PRIZEMAN. BRITISH ASSOCIATION AT WINNIPEG

EARLY in November 1906, J.J.'s pupils and friends were delighted to hear that the Nobel Prize for physics had been given to him. The formal award by the Royal Academy of Science of Sweden was 'In recognition of the great merits of his theoretical and experimental researches on the discharge of electricity through gases'. The selection committee included Professors Knut Ångström (Chairman), Hildebrandson, Hasselberg, Arrhenius and Grandquist. On the way to Stockholm, he and Mrs Thomson spent an interesting day in Copenhagen. The presentation was made by King Oscar at the great hall of the Royal Academy of Music, Stockholm, on December 10th. J.J.'s bows to the Royal party were considered to have been creditably executed—none better.

The allocution on the presentation to him was given by Prof. J.P. Klason, of Stockholm, President of the Royal Academy of Sciences. The other scientific laureates were Moissan, Golgi, and Ramon y Cajal.

At the banquet in the evening, J.J.'s health was proposed in English by Prof. Knut Ångström. He gave his Nobel Lecture before the Royal Academy of Sciences the next day. It was on 'Carriers of Negative Electricity' and covers ground which we have already traversed. Finally the laureates were entertained to dinner by the King of Sweden.

To F.V. THOMSON:

Holmleigh. Nov. 14th, 1906.

Dear Fred,

Very many thanks for your kind letter of congratulations. I am very glad to have the prize not only because it is a very substantial sum of money, but because it is a testimony to my work from foreigners who are entire strangers to me and could not have been influenced by personal considerations. It is gratifying that in three years the prize for Physics has twice been given to an Englishman. We are very busy this term at the laboratory and I shall soon have the building of the extension of it on my hands. The small extension to the house [Holmleigh] has been very successful. It is surprising how much larger the rooms look though the actual increase in size is not very large . . . .

By the bye, if the South Africans come to Manchester if I were you I should go to see them, they were here on Saturday week and I took Georgie to see them play, they are easily the best team I have ever seen in my life, their forwards are not so good as the New Zealanders, but their backs are much better, and they play an exceedingly attractive game from the spectator's point of view and are scrupulously fair.

With many thanks and much love from us all,

Your affectionate Brother, J. J. THOMSON.

To F. V. THOMSON:

Grand Hotel, Stockholm.   Dec. 10*th*, 1906.

This is a most beautiful country, and the people are most hospitable, so much so indeed that it is rather embarrassing for all callers are shown up to one's bedroom and I have just this minute had to keep a lady out who spoke no English and who could not come in because Rose was dressing.

The annual Cavendish dinner held shortly after his return was held at the Trocadero in London, to give old pupils the opportunity of attending to congratulate him. Horton was in the chair, and had J.J. on his right and Townsend on his left. It was a successful event, and graceful tributes were paid to J.J. by Townsend and others. A song entitled *An Emanation* referring to the Nobel award was written by Robb for the occasion.

In the above quoted letter to his brother, J.J. says: 'I shall soon have the building of the extension of [the laboratory] on my hands.' We noticed in a former chapter the extension of 1896 on the South side of the original laboratory. This had relieved the acute pressure of space which had arisen from the need for teaching simple practical physics to medical students. But in the meantime the difficulty had arisen at the other end of the scale. The number of research students flocking to the laboratory to work under J.J. had become so large (30 to 40) as to make the working conditions very uncomfortable. There was competition even for the space on a single table, and some who would otherwise have continued at the laboratory were driven away to work elsewhere in Cambridge, or to accept prematurely a post in another university. The Professor did not reserve a working room for himself but shared it with one or two others.

J.J. about this time received the following letter:

Terling Place. Jan. 2, 1904.

Dear Thomson,

You may have seen that I propose to give the Nobel prize to Cambridge. I am writing to ask whether you have any suggestion to make as to the form the gift should take. Physics have of course the first claim, and one naturally likes to think of something definite attained, though I believe many think that gifts are more valuable when quite unrestricted.

Please turn the matter over in your mind and answer at leisure. I shall probably consult one or two other friends.

Yours very truly, RAYLEIGH.

The prize is £7000 odd.

His reply is not extant, but he is remembered to have urged the needs of the Cavendish Laboratory, and to have insisted particularly on the difficulties which had arisen from radioactive contamination of the building, due to the liberation of gaseous radon, and the semi-permanent solid active deposit to which it gives rise. This made the more refined experiments with electroscopes and electrometers difficult to carry out, and it was very desirable to have a separate laboratory which would be kept as free from such contamination as possible.

Rayleigh ultimately assigned £5000 of his gift to the Cavendish Laboratory for a new building, the balance going to the University Library.

J.J. wrote in reply (June 21st, 1905):

Your most generous gift to the university comes most opportunely, as now we shall be able to get the site next the laboratory. I was getting to have some fear that we might have this taken from us as part of the site is to be used for an Examination room, and they proposed to come dangerously near the laboratory, but now that we can make use of the ground at once I am sure they will let us have what we want.

This site was the frontage of Free School Lane to the North of the original laboratory, and here the new extension was built at a cost of about £7000. Thomson had found £2000 from accumulated fees to add to Rayleigh's £5000. The architect was Mr W. M. Fawcett, who had designed the original laboratory. It included a large basement and a number of small rooms on the second floor for research, together with a large lecture room, a library and chemical room, and a room for the demonstrators on the first floor.

J.J. was very hard pressed with work at this time, and it may have been for this reason, or because he was not well advised, that several contretemps occurred. Thus it was found (I believe after the building had begun) that the plan would have infringed Ancient Lights. The man whose lights would be infringed refused to deal, and that part of the building had to be reduced. Then, again, no provision was made for darkening the lecture room. When Hayles, the lecture-room assistant, said to J.J., 'What are we going to do about darkening the windows?' J.J. said naïvely, 'What do we want to darken the windows for?' forgetting that a lantern was in constant use. This, however, was only a momentary reaction. If the matter had been put before him at the proper time, no doubt he would have agreed that blinds were necessary.

The new laboratory was formally opened by Lord Rayleigh on the occasion of his installation as Chancellor of the University, June 1908. J.J. and others spoke. His speech is not recorded but I remember one sentence which struck me as I listened: 'In the course of experience of many generations of students, I have known far more to fail from lack of grit and perseverance than from the want of what is commonly called cleverness.'

In connection with Lord Rayleigh's tenure of the office of Chancellor, he, like some of his predecessors, wished to have a resident correspondent in whom he could feel confidence, who would keep him *au fait* of University affairs, and he asked J.J. to undertake this office. Most of the letters written in this connection are not of general interest, but the following written in view of his approaching inauguration is an exception:

Holmleigh. May 20th, 1908.

Dear LORD RAYLEIGH,

After a great deal of hesitation I promised to communicate to you the following suggestion, in doing so I am, however, merely acting as a channel of communication between a member of the senate and the Chancellor. This afternoon Professor Waldstein came to me and asked if it would be possible to bring before your notice the name of J. G. Frazer, the anthropologist of *Golden Bough* fame as a possible recipient of an Honorary Degree. The points are that it has not been possible for the university to give him any recognition of this character before as he has until quite lately been resident here, indeed he only left this term to take up a Professorship at Liverpool, that the council were practically forced this year by the Pan-Anglican Congress to fill their list of degrees with Bishops, and that Frazer has done a very great deal for the reputation of the University. I pointed out to Waldstein that the Chancellor's nominations were entirely a matter for himself and that you had every opportunity of getting suggestions for yourself if you wished for them; he was, however, so keen on the question that I found he was determined to go to Mrs Sidgwick if I did not let you know about the matter so I thought it would save trouble if I wrote to you myself. I know Frazer well, and I am sure that the thought that he was being pushed on anyone's notice would be intolerable to him; he is in fact morbidly conscientious, he wrote to the Council of Trinity College offering to resign his Fellowship because someone detected a misquotation in the *Golden Bough,* and he insisted against the wish of the directors in paying a Railway Company £40 because he found after a long time that a bundle of books he had been in the habit of taking in the train with him daily between Edinburgh and Glasgow was above the permissible weight and after a laborious calculation of the number of journeys he came to the conclusion that he owed £40 for excess of fares which he insisted on paying:* he would not like, I am sure, anyone being troubled with his claims which I believe are exceedingly strong, and whose recognition would as a matter of fact be exceedingly popular here. I hope I have not done wrong in sending on Waldstein's opinion, but being new to my duties I thought I had better err on the side of excess of zeal in spite of the adage.

Yours very sincerely, J. J. THOMSON.

* J.J. was fond of this anecdote. As he was accustomed to tell it in conversation, he added that Frazer had actually written a stiff letter to the Directors, telling them that they had no right to refuse money which was due to their shareholders!

For many years previous to the war Lady Thomson gave an annual tea party followed by an entertainment to a Mothers' Meeting conducted by herself and Miss Graves at Barnwell. This was one of the great events of the year for the women concerned. They brought their husbands and babies. The assistance of Dr Alexander Wood and Dr F. Horton was enlisted, and they collected such of the Cavendish research students as could sing or recite, supplementing their efforts by lantern show, cinematograph or conjuror. J.J. invariably attended these treats and seemed to enjoy them. At tea he took charge of an urn at one end of a long table and filled cups which were passed round. Some manœuvring for position was often noticed among the men trying to sit next the Professor, and so have the privilege of conversation with him. He knew the names of some of them, and was able to show a surprising knowledge of things in their sphere of interest. There was indeed some anxiety lest he should get so interested in the talk as to forget that the tea urn was running. After the tea was over he produced from his bulging pockets packets of cigarettes for the men. During the entertainment he would join vigorously in the chorus of a song, such as 'Lucky Jim'. His contribution was loud and in monotone. At the end of the party he handed to each of the men a packet of tobacco as they left, and the hosts and their helpers retired somewhat exhausted to have supper at Holmleigh.

In the autumn of 1908 Thomson was gazetted a Knight Bachelor on the recommendation of Mr Asquith. I remember receiving a letter from him in answer to my congratulations, in which he said that he had had some doubt about accepting, but had been led to decide for it by the consideration that it might contribute to the success of the coming British Association Meeting at Winnipeg, over which he was to preside.

To F.V. THOMSON:

Holmleigh.   Nov. 5th, 1908.

I am not sure but that I would rather be without the title myself, but I always think it seems rather conceited to refuse such things, it implies that you think you are so distinguished that such things would make no difference to you.

I suppose that now the election is over you will be very busy in the American trade. I certainly expected Bryan to make a better fight, it must be disappointing to make fifteen speeches a day for almost six weeks and then poll fewer votes than ever.

New College, Oxford. 24th Nov. 1908.

Dear Sir Joseph,

I am glad of the opportunity of congratulating you on your recent honour, as I shall be unable to go to the Research students dinner this year and take part in the after-dinner merriment. I suppose it was owing to the inability of a Trades Union Government to distinguish between the relative merits of working men that they did not rise to something more appropriate than an ordinary knighthood.

I am sending you MacClelland's nomination certificate [for the Royal Society], which I have received from Rutherford. I hope he will get in next time. Wilson has been staying a weekend with me finishing the examination for the London University. He is thinking of going to Montreal if they make the salary good enough.

Yours sincerely, JOHN S. TOWNSEND.

To Mrs H. F. REID, *Dec.* 11*th*, 1908:

It was indeed nice to get your good wishes and your 'benighted' story. I can cap it with another. Joan has a little friend who comes to have lessons with her and whose ideas about knights are derived more from history than from reality. When she heard, she said she was quite pleased because it would be so interesting to see one walking up and down the West Road in a suit of armour.

To the same:

Holmleigh. Aug. 10*th*, 1909.

I wish you could see Joan, she is getting quite grown up and is showing decided originality for she has taken to keeping worms as pets. I have to visit them about twice a day, and they are literally unearthed for my inspection. I have also been taken to task for my Natural History for I was rash enough to say that I thought they were not a species of snake, which I am afraid is one of their attractions, however, as she is sure I am mistaken, no harm has been done.

About the time (1909) when Mr Lloyd George became conspicuous as the author of the 'People's Budget' J.J. was often taken for him in public places. Neither of them were good customers of the hairdresser's shop, and this circumstance doubtless helped the illusion. Some of Mr Lloyd George's admirers were so convinced of the identity that they asked the favour of shaking the supposed hero by the hand. 'What answer do you give?' I asked. 'Oh,' replied J.J., 'I do shake them by the hand when they ask me. I do not want to deny them a pleasure.'

We have seen that J.J. had been President of Section A of the British Association in 1896. In 1909 the time had come for him to be President of the whole association which was to meet at Winnipeg. At the same time, and very appropriately, Rutherford was to be President of Section A. This was the third meeting of the Association in Canada. The previous ones had been at Montreal in 1884 under the Presidency of Lord Rayleigh, and at Toronto in 1897 under Sir John Evans. J.J.'s personal party consisted of Lady Thomson, his son George, aged 17, and his brother Fred, who took charge of the finances, and relieved the others of all trouble on this head. They left Cambridge on August 13th and went out via Quebec, and on to Montreal. They then went southwards into the United States, visiting Niagara, Chicago, St Paul and Minneapolis, and thence to Winnipeg.

J.J.'s Presidential Address was given in the Walker Theatre. The retiring President, Sir Francis Darwin, was not present, and his place was taken by Prof. George Carey Foster.

J.J. began by speaking of the amazing growth of prosperity in Canada in general, and in Winnipeg in particular, then spoke of his predecessors, Rayleigh and Evans. He referred to the fact that British Science had been largely developed by men of independent means, like them, and hoped that more such might be attracted into this field in the future. The tendency of events has been altogether otherwise. Science has become more professional, and more dependent upon organisation and team work. The scope for independent amateur effort has become less, and however much we may regret it, such contributions to British Science are tending to become a negligible fraction of the whole. The few young scientific workers of means have found it best to work in university laboratories rather than independently.

J.J. continued with an account of his views on education, and the educational value of research. This topic is treated in Chapter XIV and may be passed over here.

He then went on to deal with his own branch of science. It may seem somewhat of a paradox, but an address of this kind given to a general audience is far more difficult to compose than a paper to a learned society. The whole English-speaking world listens, and expects to be interested, and to hear something new to it; and at the same time it is not prepared to pay more than ordinary attention, or to seek further afield for preliminary explanations; and the address has for this reason to be self-contained. It would be idle to pretend that the successive presidents have always succeeded in their task; for a pure mathematician, for example, it is an almost impossible one. J.J.'s position was by no means so difficult, for the subjects in which he was interested had much of the vigour and absence of subtlety which is the privilege of youth.

He began by running over the main results of the Cavendish Laboratory researches from the discovery of X-rays onwards, as narrated in earlier chapters. He dwelt on the contrast between positive and negative electricity, and when he came to deal with the ether, he still showed that robust confidence in its reality and necessity which was maintained up to the end by Sir Oliver Lodge. Einstein had already queried it, and had pointed out that while the conception of atoms had been helpful to the progress of science, and had become more and more definite, coherent and self-consistent as time went on, the opposite was the case with ether, which was constantly causing trouble by the fantastic and mutually contradictory properties which it was necessary to assign to it. J.J. was not an early convert to this line of thought. Certainly he was not converted at the time of this address, and not fully even at the end of his life (see p. 203).

Is the ether dense or rare? Has it a structure? Is it at rest or in motion? are some of the questions which, he says, force themselves upon us. It is not now fashionable to put these questions. No one now thinks that they will be answered in a way that will satisfy the questioner who puts them, and will lead to ultimate agreement. J.J. in this address seems to declare in favour of a dense ether, some 2000 million times that of lead. Sir Oliver

Lodge was at this time of a similar opinion. It is not now considered that this is a fruitful line of thought. It could only become so if widely different methods of measurement led pretty accurately to the same result, as they have done, for example, in determining the wave-length of light, the number of molecules in a cubic centimetre of a gas, or the constant which determines the quantum of action.

The remainder of the address dealt with the development of radioactivity, and some attention was given to its geological aspects, in connection with the internal heat of the earth and the duration of geological time.

Although the local audience cannot have included a large proportion of academically trained people and although there was some chaff in the local papers about the more difficult passages, the address went down very well, helped no doubt by J.J.'s personality. This always appealed to the 'plain man', who saw instinctively that there were no airs of superiority about him and that he was trying to be as intelligible as he could in the circumstances.

After the address Mr W. Sandford Evans, the Mayor, and Lord Strathcona spoke to thank the President for his Address, and to welcome the visitors.

Besides giving the Presidential Address, J.J. opened a discussion on Positive Electricity in Section A.

After the close of the Meeting on September 2nd, the officers of the Association and certain invited guests went on a trip by special train to the Pacific and back again by a different route. The train was a very long one and was fitted with sleeping accommodation, an observation car, and every facility for meals. At the more important stopping-places they were received by the local notabilities and entertained to a meal, and J.J. had to make a speech. The British Association party were present every time and it would have taxed most people's ingenuity to find something new to say. However, J.J. succeeded in doing so. On one occasion, he began on the subject of collecting postage stamps, which, he said, had started his great interest in Canada when he was a small boy. There were receptions at Regina, Moose Jaw, Calgary, Victoria, Vancouver and Edmonton.

Several days were spent at Vancouver City and on the Island. The party then returned via Edmonton back to Winnipeg, Toronto, Montreal and Quebec and so home.

To PROF. H. F. REID:

Quebec. Sept. 24, 1909.

We had a good deal of amusement on the trip, at one place, Moose Jaw, we were met by a brass band and escorted to a triumphal arch built up of the produce of the district. Concealed in it was a tree stump on which I had to be mounted to make a speech. My speech was, however, very short for the top of the arch was comprised of very heavy bags of flour of a brand produced in the city and of which they were very proud and as the construction of the arch left much to be desired from an engineering point of view, it shook in a very ominous manner when I got on my stump and I expected to have one of the sacks on my head every moment. I naturally therefore did not indulge in any very prolonged oration. It was very invigorating to be in the middle of the intense optimism and energy of the West, it is certainly going to be a great country, although it may not, as the Mayor of Vancouver predicted, result in Vancouver being within fifty years the biggest city in the British Empire. . . .

I have visited the laboratories of my old pupils at Minneapolis, Toronto and Montreal, and have been immensely interested. I think McLennan's laboratory at Toronto is the best I have ever seen.

J.J. enjoyed the trip immensely, especially the entry into Canada up the St Lawrence into Quebec, and later the prairies, the wide unending expanse day after day, and all night long, and the Rocky Mountains. There were old pupils to greet him everywhere with the warmest of welcomes, as was also the case on each of his visits to the U.S.A. He was an excellent sailor, and the roughest crossing did not trouble him at all.

On December 22nd, 1909, the twenty-fifth year of J.J. Thomson's tenure of the Cavendish Professorship was completed, and it was felt that something ought to be done to commemorate the occasion. As the result a history of the Cavendish Laboratory was published in 1910; it was the joint work of several authors—all of them past or present workers in the laboratory, and by their collaboration it was possible to have practically the whole story from the beginning as it appeared to eyewitnesses. J.J. himself contributed a general survey of the twenty-five years of his Professorship, the details being filled in by contributions from H. F. Newall, E. Rutherford, C. T. R. Wilson and N. R. Campbell. There was also an account by L. R. Wilberforce of the Development of the Teaching of Physics.

There was a commemoration meeting at the laboratory on November 12th, 1910, with the Vice-Chancellor in the chair. The main feature was the presentation of a bound copy of the History of the Laboratory to the Professor by Glazebrook, and the Professor's reply, in which, it is scarcely necessary to say, he talked solely of those who had worked with him and under him, and not at all of himself.

There was a conversazione afterwards with experiments on view, and this was probably the first occasion on which J.J.'s positive ray photographs (parabolas) were on view (see next chapter). I can well remember his characteristic grin of pleasure as he showed them to a group of whom I was one.

In 1910 (October 6th–12th) J.J. went with Lady Thomson to Berlin for the centenary of the University, staying with Prof. and Mrs Warburg, in one of the Helmholtz houses.

In 1910–11 he was President of the Institution of Junior Engineers and gave an address 'On the influence of pure science in Engineering'. Many of the topics were those on which his views had been expressed elsewhere, but the following passage is more distinctive:

> I have sometimes seen it argued that from the commercial point of view the wisest plan is to let other people spend their money on experimenting on inventions or developing new processes, and wait until the preliminary difficulties are overcome and success assured before taking the matter up. The idea seems to be that from a business point of view it does not matter much about the early stages of, say, the motor car or flying machine, since in these stages the machines will be so crude and dangerous that the demand will be small and the business unprofitable, that it is quite time to take the matter up when the machines have been so far perfected that a great many people are willing to use them. Then we, with this experience and our large capital and well-equipped factories, can improve the details of the machine and do a large and profitable business. I must confess that I think such a

course undignified, unsportsmanlike and quite unworthy of the great traditions of English engineering, which has always been in the forefront with pioneering work. It seems to suggest that we should let others kill the game, and after it is killed we should come in and eat it. This is certainly undignified, it will not add to our prestige, and though as a child in finance I hardly like to express an opinion on this point, I doubt if it is good business. Though the discoverers of some of the most important inventions and developers of some of the most widely used processes have, it is true, not been the ones to whom the profit has fallen, yet this has been almost always due to their lack of capital and ignorance of finance. Now that pioneering work is being done more and more by large firms with abundant capital, and knowing everything there is to be known about finance, I find it hard not to believe that under these conditions they will retain not merely the honour of the invention, but also the lion's share of any profit there is to be made out of it.

In August 1911 the members of the Institution were received at Cambridge, and a garden party given for them at Holmleigh.

A few other events deserve brief mention. In 1911 J. J. Thomson was President of the Physical Society of London. They came to Cambridge for the day on June 25th and a party was held for them at the Cavendish Laboratory. In 1913 he attended the second Solway Conference in Brussels and read a paper on the Constitution of the Atom.

In June 1914 he gave the Romanes Lecture in the Sheldonian Theatre at Oxford on The Atomic Theory. This was in connection with the Roger Bacon Commemoration.

CHAPTER IX

POSITIVE RAYS

W E have followed Thomson's achievements in the discovery of the electron—the unit of negative electricity. The years round about 1898 when this was done may be considered the highest peak of his career as an experimentalist: it was followed by a comparatively fallow period between, say, 1901 and 1906, after which followed a second peak, not so high as the former, but none the less characterised by achievements of great importance, which would in themselves have been enough to gain for his name an enduring place in science. This peak, which was gradually worked up to, may be considered to have culminated about 1912.

Thomson felt that his ideas about positive electricity had not the same definiteness and success as his ideas about negative electricity. He therefore turned to the study of positive rays as the most likely opening. In telling what he did it appears undesirable to introduce knowledge which has been gained since Thomson's active career as a man of science was over, for such knowledge is no part of the story of his life. I shall therefore ignore it, and tell the story so far as it unfolded itself to him while he was at work.

We have seen how the motion of the negatively electrified ions at low pressures in electric and magnetic fields revealed them as particles of sub-atomic dimensions, with a specific charge of about 10^8 coulombs per gram. The first to make measurements of this kind on positively electrified particles was W. Wien, of Würzburg, who in 1898 observed the magnetic and electrostatic deflection of positive rays. We shall return to the full consideration of these experiments presently. In the meantime, it will be enough to mention that the value of e/m found, though not very precise, was clearly of the order of magnitude of that found for the hydrogen atoms in electrolysis, i.e. 10^5 coulombs per gram. Thomson had apparently read this paper, and mentioned the result in his own paper of 1899 on the masses of the ions in gases at low pressure. He also mentioned orally in a paper, read before the

Cavendish Physical Society about that time, that Wien had found that e/m for positively electrified particles was of atomic magnitude: but when I went up to him afterwards and asked how this had been done, he replied, rather to my astonishment, that he did not remember! He had concentrated on the electron, and had evidently given only very superficial attention to Wien's work at that time.

His own first work (1905) on the magnetic deflection of positively electrified particles was in connection with the emission by red-hot metals. This was done by the same method that was used for the emission of negative electrons from carbon (see p. 112). The results were very capricious and uncertain, the particles sometimes refusing to be deflected even in the strongest fields that could be applied. J.J. was apparently at first inclined to put the blame for this uncertainty of behaviour on the inconstancy of the heating, which was by a transformer run from the Cambridge (local) public supply, and he frequently and vigorously demanded through the telephone that the supply voltage should be kept constant. The then manager of the Supply Company was not the man to take this lying down and he retorted by sending in a bill: 'To man's time listening to complaints on the telephone, 1s.' This lay about in the entrance lobby of the laboratory for a long time.

The general result of the investigation was that the particles were charged atoms—and often very heavy atoms. It was suspected that they were in part platinum atoms from the wire.

In none of these experiments were the various species of atoms, carrying perhaps various charges of positive electricity, exhibited separately. It was one of Thomson's great successes to have ultimately achieved this, and we must now give the necessary explanation of how he was in the end able to do it.

Goldstein in 1886 observed that if holes or channels were made in the cathode of a discharge tube, and the pressure reduced to a fraction of a millimetre, then rays with luminous tracks could be seen streaming through the holes into the space behind the cathode. These rays, called at first canalstrahlen or canal rays, though in some respects resembling the cathode rays in front of the cathode, were in other ways conspicuously different. They produced a different colour in the gas along their course, a different kind of phosphorescence of the glass, and, what was more important, they were, by comparison, quite insensitive to magnetic deflection. Goldstein considered that they could not be deflected at all, but, as we have seen, Wien, in 1898, succeeded in deflecting them by very powerful magnetic forces. The sense of the deflection was, however, the opposite of that for cathode rays, thus showing that these rays consisted of positively charged particles. It then became easier to understand something of the origin of these rays. In the dark space in front of the cathode there was a region of very strong electric force, and any positively electrified particle that might be found in this region would be attracted towards the cathode, and would acquire a high velocity, which would carry it through the channel into the region behind the cathode. This region is free from electric force, and the particle moves on like a bullet when it has left the muzzle of the gun, receiving no further impulsion. It has acquired a velocity not greater than that which the voltage drop in the dark space can give it, but often less: for the particle may start in any part of the dark space, and may experience less than the full potential drop.

Since the rays had been shown by Wien's experiments to carry a positive charge, Thomson proposed to drop the name canal rays and to call them positive rays. They will be called so here.

Wien's experiment showed very clearly that complicated processes were going on in the space behind the cathode traversed by the positive rays. He found, for instance, that if these rays passed down a long tube, and a strong magnetic force was applied locally, while some of the rays were deflected sideways, others passed on unaffected. These latter had evidently lost the positive electric charges which they had originally possessed, and which had enabled them to acquire velocity while they were approaching the cathode. The magnetic field separated those rays which had a charge from those which had none. But it was found that if a long tube was used, and another local magnetic field was applied farther on, a fresh crop of deflectible rays had made its appearance, or in other words, rays which were temporarily neutral had become charged again. In this way, Wien inferred that the rays were frequently passing from the charged to the uncharged state in the course of their passage along the tube. Thomson also independently reached the same conclusion by similar methods, but Wien's experiments were the first.

It is evident that while the character of the particles is constantly changing in this manner, no satisfactory examination of them can be made, and in fact the measurements of Wien on the electrostatic and magnetic deviation only gave the vague general indication that they were charged atoms. Thomson did not at first succeed in doing very much more, but, by a combined process of experiment and reflection, he ultimately worked out the essentials of a successful technique. The discharging and recharging process depends on the rays ionising the residual gas, and thereby producing electrons which can neutralise the positive charge, or can be shed off again at collisions. In the absence of residual gas this will not happen. But here comes a difficulty, which might, and at first no doubt did, seem almost insuperable: for if the gas pressure is evanescent, it becomes impossible to pass the discharge which is required to generate the positive rays.

Thomson mitigated this difficulty to a certain extent by using a very large discharge vessel, which allows the discharge to pass with comparative ease, even when the pressure is low: but this, though a useful step, was not enough. A further improvement gradually came into view. In order to define the pencil of rays accurately, a very narrow channel in the cathode became desirable, and hypodermic needles were used at one stage. Later, even narrower tubes, only a tenth of a millimetre in diameter and as much as seven centimetres long, were substituted. These were made by drawing down copper tubing to the desired diameter and straightened by rolling between plane surfaces. The use of these long narrow tubes opened up another possibility. Pressure is only equalised very slowly through such a narrow channel. A suitable discharge pressure could be maintained for generating the rays (say $\frac{1}{100}$th mm. of mercury), and by continuous exhaustion of the observation space the pressure then could be kept very much lower (say $\frac{1}{1000}$th mm.), and the secondary phenomena of losing and gaining charges much reduced in importance.

It would scarcely have been practicable to do this if no better method of exhaustion had been available than the old Töpler pumps used up to near that time, for these pumps are not effective to remove vapours or condensible gases. But Dewar had by now introduced his method of exhaustion by means of charcoal cooled in liquid air, which most effectively takes up all gases except hydrogen, helium and neon. The laboratory had acquired a small liquid air plant, and to work it a gas engine of 10 h.p. had been put in instead of the old one of 2 h.p., which had served for driving the dynamo and the lathes and other machines in the workshop. The liquid-air plant was the gift of T. C. Fitzpatrick, afterwards President of Queens' College, whose name has already been mentioned, and rarely has money been better spent. The possibility thus gained of using Dewar's method made the whole difference to the future progress of Thomson's investigations.

In fact, I remember J.J. saying: 'I never saw a parabola* until I put on the charcoal and liquid air exhaustion.'

When this method first came into use at the Cavendish, J.J. pointed out to Everett that he might perhaps obtain a still lower temperature and better absorption by causing the liquid to boil under a reduced pressure by pumping off the evaporated gas. Everett considered this advice and saw an easier method of getting what he considered was the desired effect. That evening J.J. found him standing over the vessel of liquid air which was cooling a charcoal-filled tube and endeavouring to hasten the evacuation of a connected apparatus by dropping small pieces of india rubber into the liquid so as to make it boil! J.J. showed remarkable restraint, and quietly explained that he was wasting the liquid air.

However, to return to the main topic. The rays passing through the fine tubular aperture in the cathode passed between metal plates between which an electrostatic field was maintained by connection with a battery of storage cells. A powerful electromagnet was arranged outside so that these plates were practically coincident with its poles; thus the electrostatic and magnetic fields were practically coterminous, the lines of force being horizontal. The theory of the experiment is just the same as that of the electro-

* See below, p. 171.

static and magnetic deflection of cathode rays (p. 87), but in this case the two deflections are observed simultaneously, the electrostatic deflection, along the lines of electric force, being horizontal, and the magnetic deflection, perpendicular to the lines of force, being vertical. Suppose now that the rays are received on a fluorescent screen, and suppose that charged atoms of one definite value of e/m, but of various velocities, are present in the rays. Those with the smaller velocities will be most deflected, and the deflected rays will therefore exhibit themselves as a luminous curve on the screen. Taking the position of the undeflected rays as the co-ordinate origin, the electrostatic deflection as abscissa, and the magnetic deflection as ordinate, the curve will lie wholly in one quadrant, say the N.E. Now what ought the shape of this curve to be? Since the abscissa of any point is inversely as the velocity, and the ordinate inversely as its square, it follows that the curve is a parabola, with its axis along the axis of abscissae and its vertex at the origin.

So far we have supposed that only one kind of particle was present. If there is a second kind of particle with a different value of e/m another independent parabola should appear, with of course the same axis and vertex, but the curve itself either outside or inside the previous one. If the mass alone is greater, the deflection will be less, and the second parabola will be inside. If the charge alone is greater the deflection will be greater, and the second parabola will be outside.

This in fact was what J. J. Thomson observed. In any ordinary case, where no extreme precautions are taken to secure purity of gas, a number of parabolas are obtained, each corresponding to one kind of particle (atom or molecule) carrying one particular electric charge.

So far we have spoken of a fluorescent screen, and Thomson's earlier work was carried out in this way, using a screen coated with the mineral Willemite, which he found by exhaustive trials to serve better than any other material which suggested itself. But one of the great improvements in technique which Thomson made was to introduce a photographic plate inside the vacuum to receive the rays. This is far more sensitive to faint details than the Willemite screen and produces a permanent record which can be studied at leisure. The advantage is very much the same as that which astronomers have found in most fields of sidereal astronomy, where photography has almost driven out direct visual observation. In the case of positive rays, photography could hardly have been used without modern methods of exhaustion, which, as we have explained, were necessary for other reasons as well. The labour of remaking the vacuum after each plate was exposed would have been almost prohibitive, if only the older methods had been available. It was necessary to retain the Willemite screen for preliminary observations, just as the ordinary photographer uses the ground glass or other form of view-finder. As explained, only half the parabola is obtained at each exposure, e.g. that with positive values for both co-ordinates. By reversing the magnetic field, the other half of the curve with negative ordinates can be obtained, and in many of Thomson's photographs this was done.

So far we have not said much about the quantitative discussion of the photographs. The actual linear deflections—magnetic and electrostatic—for a point on one of the parabolas can be measured on the photographs, and they could then be turned into angular measure from the known dimensions of the apparatus. It is further necessary to know the absolute value and distribution of both the electrostatic and magnetic fields. The general theory is the same as for the cathode rays, as explained on p. 87. We do not here go into the more technical aspects, such for instance as the corrections to be made if these fields are not uniform. When all this has been carried out, we have the data for finding e/m, and the result is to identify one of the parabolas which appears on all the photographs as due to particles having the same value of e/m as the hydrogen atoms in electrolysis, namely 96,000 coulombs per gram. It is concluded that these particles are hydrogen atoms carrying one atomic charge, though there is, it is true, the alternative possibility of hydrogen molecules carrying two atomic charges. This, however, can be excluded by other considerations upon which we do not here enter. This parabola having been identified, the plate is as it were calibrated, and the measurement of any other parabola can be interpreted, and the value of e/m determined by comparison. It is then found that the other parabolas allow of a simple interpretation. They can be accounted for by atoms or groups of atoms which are likely to be present in the tube, or which have been intentionally introduced, carrying one or more elementary charges. Thus, if we exhaust a tube originally containing air, nitrogen atoms are not conspicuous, the parabolas being those due to atoms and molecules of hydrogen, atoms of carbon and of oxygen, molecules of CO and CO_2, and atoms of mercury, the last due to mercury vapour from the pump. The original source of carbon and hydrogen is probably water vapour and carbon dioxide from the walls of the tube, but these are decomposed or recombined in a different way by the action of the discharge.

When hydrocarbons are introduced into the tube, in addition to the parabola corresponding to the carbon atom, those corresponding to CH, CH_2, CH_3 make their appearance, showing that, although these compounds are not recognised in chemistry, and are not obtainable in bulk, they can none the less have an independent existence. The same is true of hydroxyl, OH. Though mercury vapour normally consists of single atoms, a parabola due to ai-atomic mercury was found, showing m/e 400 times the value for the hydrogen atom. Another interesting point was that seven different parabolas attributable to the mercury atom were found, this atom being able to carry charges of 1, 2, 3, 4, 5, 6 or 7 times the unit electronic charge.

But perhaps the most important result of Thomson's positive-ray experiments was in his detection of two separate parabolas due to the rare gas neon. Neon was not at that time a commercial article, as it has since become owing to the development of neon signs. Thomson got his supply from Dewar, his Royal Institution colleague. Dewar's method of obtaining a residue rich in neon by the treatment of atmospheric air by cooled charcoal made this gas more easily available than had been the case previously. Thomson got two parabolas from neon, indicating atomic weights of 20 and 22, the latter being much the fainter. The 'chemical' atomic weight is 20·183, derived from the original density measurements of Ramsay and Travers. It now seemed probable that neon was a mixture of two gases of atomic weights 20 and 22. In the field of radioactivity groups of elements of different atomic weight were known which were chemically indistinguishable. They had been called isotopes by Soddy, who was the first to recognise them. It now seemed that neon was a mixture of two isotopes.

Although Thomson was the first to observe the resolution of neon in this way, and although he suggested the true explanation, he was never very emphatic about it. He did not adopt the view that the heavier constituent was a compound NeH_2, which could have given the observed atomic weight within the then limits of experimental error: but he does not seem to have been convinced, at all events up to the time that his own experiments were published, that this explanation was absolutely excluded. The later work of Aston, proving that it could not be so, does not fall within our subject. Thomson's method, although it clearly showed the difference between the parabolas 20 and 22, was not developed by him to a high enough resolving power to distinguish between 20 and the value of about 20·2 derived from the density.*

Dr F. W. Aston succeeded G. W. C. Kaye as Royal Institution assistant in 1910. He had come from Birmingham, and was recommended by Thomson's great friend, Poynting. He carried out much of the work on positive rays from 1910 to 1913. The following from Dr Aston's appreciation in *The Times* (September 4th, 1940) is too good to be lost:

Working under him never lacked thrills. When results were coming out well his boundless, indeed childlike, enthusiasm was contagious and occasionally embarrassing. Negatives just developed had actually to be hidden away for fear he would handle them while they were still wet. Yet when hitches occurred, and the exasperating vagaries of an apparatus had reduced the man who had designed, built and worked with it to baffled despair, along would shuffle this remarkable being, who, after cogitating in a characteristic attitude over his funny

* This was achieved by Aston's focusing method, and also later by Zeeman's improvement of Thomson's own parabola method. J.J. seems to have always been haunted by this suspicion about hydrogen compounds, and for this reason hesitated for a time to accept Aston's later results about isotopes of other elements. He took a rather curiously hostile attitude about it at a discussion on the subject at the Royal Society in 1921.

J.J. Thomson in 1912

old desk in the corner, and jotting down a few figures and formulae in his tidy handwriting, on the back of somebody's Fellowship thesis, or on an old envelope, or even the laboratory cheque book, would produce a luminous suggestion, like a rabbit out of a hat, not only revealing the cause of trouble, but also the means of cure. This intuitive ability to comprehend the inner working of intricate apparatus without the trouble of handling it appeared to me then, and still appears to me now, as something verging on the miraculous, the hall-mark of a great genius.

The great merit of Thomson's experiments on positive rays is that they gave a new method of separating different kinds of atoms and molecules, and determining atomic and molecular weights, entirely independent of those· traditional methods which had been developed by chemistry. The result was that new kinds of atomic groupings were revealed which were not able to survive long enough for the traditional methods to be applied. They also gave much direct information about the electrical charges, positive and negative, which atoms and atom-groups can take up. The full bearing of this is perhaps not yet apparent. Further, and most important of all, they gave the clue to the existence of sub-species or isotopes of the ordinary non-radioactive atoms which the traditional methods of chemistry had, up to that time, wholly failed to reveal.

With the investigations on positive rays the most important part of Thomson's career as an experimentalist was over. He did indeed continue to work in the laboratory and some further experimental investigations were published, but they were not comparable in importance with the earlier work, and to dwell upon them in detail would be somewhat of an anticlimax.

CHAPTER X

IN WAR TIME

To J.J. Thomson as to other parents of men of military age, the war of 1914–18 brought grave personal anxieties, anxieties which he felt very deeply, dreading every ring of the door bell, for fear that it might bring the fatal telegram.

His mind did not usually run on questions of clothing, but all the more the following letter throws a flood of light on what primarily occupied his thoughts.

To G. P. THOMSON:

Cavendish Laboratory. Oct. 14th, 1914.

I hope if you can hear of anything, whether of the nature of clothes or appliances which would increase your comfort in the field, you will buy them and let me pay for them. I don't think money could be better spent. I should think there must be coats made which are warm and waterproof at the same time, could you not write to some of the army outfitters and find out? I see a Surgeon Major in to-day's *Morning Post* advocates the use of waterproof linen trousers such I suppose as motor-cyclists wear; these could not weigh much. I see too that they all advocate boots which are plenty big enough. I have no doubt you have seen these but what I want you to know is that I am anxious that you should not hesitate on the score of expense to buy anything that might be useful or make you more comfortable. I should be very glad to pay. I am writing this while Everett is trying to stop a leak which interrupted a very interesting experiment.

To G. P. THOMSON:

Nov. 19th, 1914. *Cavendish Laboratory.*

It looks as if we might have quite a contingent of the wireless at the Laboratory, we are at Lefroy's suggestion investigating that new hot wire receiver, and Wright has come down to work at it, he wanted to bring a company of the wireless who are at present in Scotland, to Cambridge, so that it may have a proper trial in the field. No one is allowed to have any private wireless installation so that we cannot test our improvements in the field unless the army gives us some help in the matter. There have been letters in the local papers about Professors of German birth living on a hill at Shelford. I suppose they refer to Gadow. Joan keeps saying how dreadful it would be if Uncle Hans were arrested, in a way that shows she thinks what an excellent joke it would be if he were.

Mysterious lights have been reported on the Gog-Magogs, which so excited the County Council that they passed a resolution asking the Home Office to take drastic steps. On investigation it turns out that these mysterious lights which appear and disappear are caused by Dr ... turning on and off the lights in his study. I saw a company of elderly M.A.'s being drilled to-day—they were bad but not so bad as their instructor who would have been helpless if it had not been for the aid of a spectator who fortunately knew his drill.

To G.P. THOMSON, then at the front as 2nd Lieutenant, Royal West Surrey Regiment:

Nov. 25th, 1914. Holmleigh.

I read in the papers graphic accounts of the sloppiness of the trenches, and the difficulty of keeping the feet warm: it struck me that the wooden sabots such as are worn by the peasants in France and Belgium might be a help worn outside the ordinary boot and stuffed with a little straw. They are very warm, and though not good for racing in they are easily kicked off. I should think they ought to be procurable in Flanders, as if I remember right the Flemish peasants wear them. . . .

If there is anything that you think you may want (not merely that you do want) please let us know and we will send it out, it takes, I believe, some time for parcels to reach you, so that if you wait until you need them you may suffer discomfort. Even if you do not need them no great harm is done as you can throw or give them away and so not load up your kit.

Write as often as you can, and take as much care as is possible of your health, you know it is a terrible time for your mother and myself. . . .

Holmleigh, West Road, Cambridge. Dec. 13th, 1914.

I saw in Macintosh's shop yesterday what seemed a very serviceable knife for military purposes so I am sending it out to you in case it should be better than the one you have. . . .

One of the men in the Laboratory has an American paper with a reproduction of a snap shot of the *Audacious** just after she had been

* H.M.S. *Audacious* had been sunk and the Government had for a time withheld the news, though it became widely known notwithstanding.

mined. The newspaper people tell me that the number and variety of the rumours they get about her is almost as great as those they got about the mythical Russians. I believe the official answer to any inquiry as to a rumour is that it is an Audacious story.

Holmleigh. Dec. 24th, 1914.

[The Military] have commandeered the big medical room at the Laboratory, and 125 men are to be billeted there. Two officers of the R.F.A. are to come here, it is rather a nuisance as there is no Officers' mess and they will have to have all their meals with us.

Jan. 7, 1915. Holmleigh.

The officers who are to be billeted on us have not yet turned up, nor have the 125 men who are to be billeted at the Laboratory. The military methods are rather exasperating, they come and say everything must be ready in a few hours and when you put everything else on one side and by a great push get ready, you do not hear anything more about it from them for two or three weeks.

Holmleigh. Jan. 19th, 1915.

The men billeted at the Laboratory do not give us any trouble: other people are not so fortunate.

Jan. 24th, 1915.

Practically all the Research students at the Laboratory have now joined the O.T.C. and so cannot work in the afternoons. I am making arrangements for them to work in the evenings instead.

To F. HORTON:

Nov. 19th, 1914.

The war has made a great difference to the Laboratory, so many are away serving in one capacity or another: we are making experiments too at the request of the War Office on a hot wire receiver for wireless messages, Wright and Ogden have got the work in hand. George has gone to the front so we are having a very anxious time.

A little later the laboratory was turned over entirely to war work, the workshop being employed in making gauges.

Previous to the outbreak of war in August 1914 it had been very little realised by the services how many new problems would present themselves which would require scientific help. Wireless telegraphy, aviation by aircraft both lighter and heavier than air, submarine detection and destruction had none of them been touched during the South African War in 1900, and although there were a few enterprising officers in the navy who were capable of dealing with such matters, it was felt by the Government and the public that they were not as much in touch with the resources of science as was desirable. In particular, there was an uneasy feeling that valuable ideas sent in from outside sources were met by an attitude of official obstructiveness, and rejected without being examined. In order to deal with this situation the First Lord of the Admiralty (A.J. Balfour) announced in July 1915 that a 'Board of Invention and Research' (short title B.I.R.) would be set up. The members of the main committee were Lord Fisher (Chairman), Sir Joseph Thomson, Sir Charles Parsons, and Dr G.T. Beilby. Associated with these were a 'panel' of consultants; these were H.B. Baker, W.H. Bragg, H.C.H. Carpenter, Sir William Crookes, W. Duddell, P.F. Frankland, B. Hopkinson, Sir O.J. Lodge, W.J. Pope, Sir E. Rutherford, G.G. Stoney and the present writer. Others were added later.

The position of this Board relatively to the Admiralty organisation undoubtedly had elements of weakness. Lord Fisher had recently left the post of First Sea Lord owing to differences with Mr Churchill, the former First Lord, who had also left. Many antagonisms undoubtedly existed between Fisher and others who remained at the Admiralty, and it may be suspected that there were some who were not willing to see the success of an organisation headed by him. Be that as it may, the Board was not effectively an Admiralty department, and its findings so far as I at least could observe did not usually carry much weight. They were apt to be pigeon-holed and nothing done. Balfour had hoped that Fisher's energy and vigour would outweigh the disadvantages of personal antagonisms, and of his inordinate love of getting even with those who in any way opposed him.

J.J. had first been brought into contact with Lord Fisher when the latter received an honorary degree at Cambridge in 1908. He made an after-luncheon speech in his usual exuberant style about our naval preparedness, 'sleep quietly in your beds' and the like, and J.J. said afterwards that it was far from giving him confidence. However, on further acquaintance with Fisher he seems to have put this feeling aside as others had done who saw him at work, and he seems ultimately to have come to regard him with respect and admiration, and even to have rather hung upon his words.

So far as I knew at the time, or have been able to learn since, J.J. did not himself develop any effective war device, though there is mention of a design of his for a non-contact mine, and for an explosion pressure-recording device based on the use of a piezo-electric crystal with a cathode-ray oscillograph. These were late on in the war, and it does not appear that the contact mine was pursued to a successful conclusion, though the pressure recorder has a place in the history of such devices.

The direction in which his help was of most value was in finding the right man for any job: because a large proportion of those who were likely to be of use had passed through his hands at the Cavendish Laboratory. Some of his other activities at the B.I.R. are described in the extracts from his letters which are given below. As will be gathered from them, the suggestions which came in from the outside public were not usually of much value after all.

To G. P. THOMSON:

Holmleigh. July 21*st*, 1915.

I have just written one letter to a man with a perpetual motion machine, and the other to a charwoman who was much upset by a bad smell and thinks it might be bottled up and used against the Germans.

To PROF. H. F. REID:

Holmleigh. Sept. 5*th*, 1915.

I have been quite busy with war work for some time now, besides getting the inventions wanted for the Navy and Army put into the hands of the people most likely to work them out successfully.

I have often as a kind of side show to interview people who have schemes which they assert will end the war in a few days, and which they will communicate if the Government will give them ten millions or so. It is quite an amusing study in human nature to sit through these interviews, there is the man with the idea, then the man who is finding the money for him to work it out, often a small tradesman with a great idea of his own importance, and of the deference due to him from his colleagues. They often bring with them a low-class attorney to see that they are not cheated, and sometimes a man with the 'gift of the gab' who has made them believe that he is the bosom friend of half the peerage and able to pull all the political wires. Some of the schemes are wild enough, one was to train large numbers of cormorants to peck the mortar from between bricks and then let them loose near Essen so that they might peck the mortar from Krupp's chimneys and so bring them down.

I hope you and Mrs Reid are having as happy and calm a time as is possible in these troublous days, let us hope that it will not be long before we can look back on these times as a kind of nightmare. I am afraid, however, that the world and life will never in our time be quite the same again.

To G. P. THOMSON:

Holmleigh. Sept. 19*th,* 1915.

I have been busy with work for the Board of Invention and Research. The result of the zeppelin raid has been that we have had since on an average about 500 letters a day suggesting plans for catching zeppelins; one was to have a thick rope heavily smeared with birdlime hung from a captive balloon.

Fisher no doubt soon discovered that the work of the Board did not 'cut much ice' at the Admiralty. There was put up in large lettering in his room a quotation, the source of which I have not been able to identify, but which ran somewhat thus: 'Mr Burke said he had little faith in any scheme where the conception was divorced from the execution', and this probably expresses his own lack of faith in the B.I.R., after he had had a little experience of its working. J.J. was heard to complain that the Admiralty did not readily give information on the simplest question, such as the kind of steel used in making mines.

The most successful work done by the Board was in the development of anti-submarine listening devices, which was carried out at Hawkscraig and afterwards at Harwich under Sir William Bragg. The staff under him consisted partly of scientific workers and partly of mechanics. When it was suggested that the pay of the former ought to be definitely the higher, J.J. did not seem to share this view. Probably he was influenced by the relative rates of payment of the assistants and the junior demonstrators at the Cavendish Laboratory.

When Lord Fisher was present, he seems to have done most of the talking. In his not infrequent absences, J.J. took his place, and became the chief speaker. Sir Charles Parsons, who had definite views of his own, but was very inarticulate, resented this strongly, and actually worked himself up into believing that J.J. was bent upon suppressing and humiliating him: nothing anyone could say had the slightest effect in mitigating this wild idea, which was the more pathetic because J.J. had in fact a great admiration for him, and was quite unconscious that anything was wrong. The following, found in J.J.'s handwriting and apparently a draft for an obituary notice, makes this clear enough.

Sir Charles Parsons threw himself with characteristic wholehearted- ness into this work [B.I.R.]. He was tireless in suggesting new methods, in giving opportunities for experiments on a large scale, and in helping by his criticism and advice those who were engaged in carrying out the investigations suggested by the Board.

He was a very agreeable as well as most efficient colleague, and though the questions before the Committee were such as to give room for wide differences of opinion, I have rarely been on a Board where there was less friction and where the proceedings were more harmonious. It was most interesting to work with him, he had the engineering instinct more fully developed than anyone I ever met; though he had taken a high place in the Mathematical Tripos he made very little use of mathematics in developing his ideas, he seemed to be able to carry these by intuition to the stage where they could be tested by the large scale experiments which his soul loved.

Fortunately all came right in the end. J.J., hospitable as usual and quite unaware that there was any breach to be healed, asked Parsons to stay for the week-end at Trinity. The invitation was accepted, and all was well. J.J. in his turn went afterwards to stay with Parsons in Northumberland.

Thomson undoubtedly enjoyed the experience of the B.I.R., which brought him into contact with much that was novel and interesting and entirely outside the academic atmosphere in which he had lived up till then. On the other hand, I do not think that the time he spent upon it was particularly fruitful as regards ultimate results.

To G. P. THOMSON:

Cavendish Laboratory. Feb. 14th, 1916.

You may have seen in the papers attacks on Lord Fisher—a good many people would like to see him back at the Admiralty where things want livening up, there are also many people who *dont* want him back, and know the first thing he would do would be to kick them out— hence these articles.

If the members of the Cabinet could hear the opinions he sometimes expresses at our Board on the War, I am not sure they would be satisfied with his appreciation of their efforts.

Holmleigh. Nov. 8th, 1916.

Lord Fisher is at present very much on the rampage because the Fleet is not fighting enough, it is a mercy that there are no reporters present at our meetings.

Ferne, Donhead, Salisbury.

Dear Sir Joseph,

It was mean of me to leave you in the lurch last Thursday, but I felt very seedy and was sure it was best to get to the above address, 800 feet above the sea and food as plenteous and good as with Joseph in Egypt! Merz* told me before leaving that his one object was to be in accord with the central committee. So I hope all you desired was agreed by him and Sir Eric Geddes† said the same! If you wish me to do anything or to see you I'll come up any day if you'll tell Phillips and he will telephone to me.

I am now engaged on the 'Meditations of Lord Fisher' and I'm going to print one hundred copies tied up in white satin ribbon full of libellous matter and I mean a copy for you printed in Great Primer type, so you wont strain your eyes any more than your mind in reading them. I've begun my life backwards and will issue it in portions till I arrive at my weaning and my birth.

Yours, FISHER.

28. 1. 18.

---

* Charles Merz, Director of Research, Admiralty, 1916–18.
† First Lord of the Admiralty.

To G. P. Thomson:

Trinity Lodge. Sept. 1st, 1918.

I have not heard any details about poor Hopkinson's accident;* it was singular that within a day it was the anniversary of the death of his father and his brother and sisters who were killed on the Alps. I remember we were staying at Aberdovey at the time and I had just finished reading the *Manchester Guardian* and as I was putting it away my eye caught a paragraph headed 'Death of Dr John Hopkinson'. I had missed it when reading the paper, exactly the same thing happened this time. I had not noticed anything about Hopkinson and it was only when I was throwing the *Guardian* on the floor that my eye caught the paragraph about his death. It is a dreadful thing, and I hardly know how they will get on without him. He had done such excellent work and had so much on his hands. I think there ought to be in the air service the same understanding that there is in the army, that those who are responsible for the policy, the brains, and the administration should not run any needless risks; it would be thought inexcusable if except under quite exceptional circumstances people like Haig or Foch or even much humbler people were to go under fire, an accident to them may result in a failure which will imperil the lives of thousands of others. I think it would be well if there were an order in the air service to the same effect. Hopkinson used to use aeroplanes almost like cabs, if he wanted to go anywhere he would fly if possible, it was very sporting and plucky but it was bad for the interests of the country. It is what they call the irony of fate that just at the time when peace really seems to be not very far off he should be killed and not live to see the full fruits of his work.

J.J.'s brother, Frederick Vernon Thomson, has already been mentioned. He had a useful but undistinguished business career with Claflin and Co., calico merchants in Manchester, who had a large connection in the U.S.A. He was one of the founders of the Hugh Oldham Lads' Club, where he spent at least two nights weekly all his life, until his mother's death in 1902, when he went every night. Some three thousand lads had been members of the club, and he was said to know them all, and at the time of his death was in correspondence with any number of them at the war and elsewhere. He organised a camp for the boys once a year, and

* Bertram Hopkinson, F.R.S., Professor of Mechanism and Applied Mechanics at Cambridge, 1903–18.

taught many of them to swim and play chess, for he was good at these himself. It was easy to see that a University career might not have suited him. He was a lovable personality, much attached to his brother and proud of his intellectual success. The brothers were as unlike as could be imagined.

To F. V. THOMSON:

Holmleigh. July 25th, 1914.

My dear Fred,

I was deeply grieved to hear of your illness, and that you are having those painful attacks again. I think when they find out what it is really due to they will be able to cure you. . . . I hope you will let me come over so that I can see the specialist and hear what he has to say. You and I ought not to be far apart when one of us is ill. . . . Rose as you will be sure was deeply concerned when I showed her your letter. She suggested that you should come here to be nursed when you were well again, that seems to me an excellent plan, do try and see if you cannot arrange it, we should then be near each other.

With our warmest love and sympathy,

Your affectionate Brother, JOE.

The anxiety which may be seen in this letter was only too well justified. A serious operation had to be performed, and Frederick Thomson spent the last three years of his life at Cambridge, where he took a small house. He died in July 1917, and representatives of the Club for which he did so much came to his funeral. He still lives as an affectionate memory with them. His brother always personally carried a wreath to his grave at Christmas for 21 years, after which he was no longer physically able to do so.

CHAPTER XI

PRESIDENT OF THE ROYAL SOCIETY

THE Presidency of the Royal Society was due to become vacant in November 1913. It had been thought likely that Sir George Darwin would be called upon to occupy it, but his premature death in 1912 led to a general wish for Thomson as President. It is curious that the two letters which survive among his papers pressing him in this sense should have come from the two men with whom there was most friction when (on a later vacancy) he did occupy the Presidency.

From SIR JAMES DEWAR:

Royal Institution. June 10th, 1913.

When I read your letter this morning my heart sank within me, I had hoped to live long enough to see you in the position of President, the only legitimate successor of Kelvin and Lister together with all their honours. My reason for referring you to the course Davy adopted was because of the fine work he continued during his term of office. Davy was far too clever to burden himself with the humdrum duties the old and fatuous Presidents we have seen lately impart to the duty. No, he considered it an honour to preside from time to time (in court dress) and left the routine work to others. Although we would not encourage court dress nowadays, still I assert Davy's view of the position was the correct and proper one, to go on with his scientific work while decorating the name of British Science. You can if you like succeed in doing what Davy did, and at no greater age than Huxley, Hooker, Wollaston and Spottiswoode, when they held office. I hope you will still reconsider the offer that will undoubtedly be made. The Society needs the help of its best and you alone can succeed to meet the wishes of all the members.

55 Granville Park, Lewisham, London, S.E.
June 14th, 1913.

My dear Thomson,

I hope you will permit me to add my entreaties to those of other friends that you will consider favourably any proposal that may be made to you to lead the Royal Society. The present is a most critical juncture in the interests of science. Something must be done to rehabilitate the Society and make it a going concern, not the sleepy hollow it is bound to be so long as it remains in the hands of octogenarian Presidents and a purely autocratic machine. Something must be done to improve the Council and make it effective against the official ring and to use the forces of the Society which are now unorganised and inoperative. You alone, at the moment, could lead such a movement, and you could best do it by looking on with approval. What I mean is that you need not give up much of your own time. Geikie's great mistake has been that he has intervened everywhere and ruined all the committees by acting as Chairman. I can appreciate your desire to confine your activity to scientific work, but after all unless some of us take a hand in the regulation of affairs things must fall into the hands of schemers who care for little but their own ends.

Yours very truly, HENRY E. ARMSTRONG.

What J.J. thought of this letter is not on record. He may have been somewhat surprised to receive it, because Armstrong's attitude towards his school of scientific thought had been far from enthusiastic, while J.J. on his side was by no means impressed by Armstrong's criticisms, as I know from many well-remembered conversations.

The Royal Society is organised with five officers, forming a kind of inner council, and a council. To invite the President to strengthen the council, so that they may be effective against the 'official ring', i.e. the officers of which he himself is the chief, seems a strange proposal. As to the schemers referred to, it is difficult to imagine what was meant.

J.J., however, did not accept the Presidency at this time. He happened to ask me casually if I had any views about who should be President, and when I said that I supposed he knew that many of the fellows wanted *him* to accept it, he said that he had refused. 'I had much rather enjoy myself doing my work here', he said, 'than be constantly going up to London to attend dinners.'

Eventually Sir William Crookes was chosen. His scientific distinction was undeniable, but he was at that time eighty-one years of age, and it was thought well to alter the usual convention of a five years' tenure to two or three years in his case. Consequently, the question came up again in 1915, and Thomson was approached once more.

From SIR ARTHUR SCHUSTER:

Yeldall, Twyford, Berks.   July 4, 1915.

The Council of the Royal Society has asked me to write to you confidentially to ask if you would allow yourself to be nominated for the Presidency of the Society next November. I may say that the wish of the Council was quite unanimous.

Personally I may add that you would be rendering a public service not only to the Society but also to the country if you would allow yourself to be nominated, and that during the time I shall continue to act as secretary, I will do everything in my power to save your time.

Holmleigh, West Road, Cambridge, July 5*th*, 1915.

Dear Schuster,

I feel it is a very grave responsibility as well as a great honour to be President of the Royal Society, but if the Council think I can be of any service to the Society in that capacity I am quite at their disposal and I feel very grateful to them for thinking I am worthy of consideration in such a connection.

It would be a very great pleasure indeed to me personally to become your colleague. It would recall the time now alas! forty-three years ago when we worked side by side in Balfour Stewart's laboratory at the Owens College. You helped me then, I am sure you will do so now.

Yours very sincerely, J. J. THOMSON.

In accepting this time he was perhaps influenced by the consideration that his own scientific work was in any case interrupted by the war, and that there were not many dinners to attend in war time. Not that he disliked public dinners, quite the contrary, but he had not been prepared to give up a large part of his life to attending them.

As President Thomson did not attempt to dominate the Council Meetings. When controversial questions were under discussion, he let his own view be known, though briefly, and without undue insistence. At the ordinary Scientific Meetings of the Society, he was admirable. In modern times, with science becoming more and more specialised, the discussions on papers read tend to become less and less general, simply because there are not enough people present who feel that they are sufficiently *au fait* of the subject to commit themselves. This is apt to have a very damping effect on young authors, who always over-estimate the capacity of their audience to grasp at short notice and in quick time ideas which they themselves have slowly and painfully evolved. Thomson was ready to throw himself into the breach, and found something to say to fill up the void, even on biological topics of which he necessarily knew very little. Possibly what he said would not have always stood critical examination, but that did not matter. It helped to make the meeting go, and the young author felt that what he had to say had not failed to interest the President. It was another example of the quality which had done so much to promote the success of the Cavendish Laboratory school in earlier years.

To SIR ARTHUR SCHUSTER:

Holmleigh. Dec. 13*th,* 1915.

I am quite of your opinion as to the necessity of having a clear understanding as to the work for which each Secretary is responsible and also for having each secretary to carry out the work for which he is responsible in the way he thinks best. Hardy* is an excellent fellow burning with desire to do as much work as possible for the Royal Society, but he seems to me to be rather in the stage when if he finds he has a spare hour or so he thinks he must be doing something energetic for the Society and goes off and does it even if he has not thought it out fully or consulted with the other officers. He will I think soon get over this and when it is directed into proper channels his energy will be all to the good. I had a long talk with him yesterday afternoon primarily about the Cambridge list, but it drifted on into the state of affairs in the office and I thought over the matter after he had gone and determined to write to you to-day about it. It seems on Friday Hardy and Harrison† had something of a tiff. Hardy said he saw now he was quite in the wrong and was going to tell Harrison so to-day. It had apparently something to do with Hardy telling some of the clerks to send out circulars about the supply of proto-zoologists and Harrison complained that it would prevent them doing other work which it was necessary to have done by a certain time. I can sympathise with Harrison as he must find it very difficult to arrange the office work when subject to interruptions like this. I should like to have a meeting to discuss the arrangements in the office. You know and I do not the relations between Harrison and the rest of the staff and so will be able

* Sir W. B. Hardy, Secretary of the Society, 1915–25.
† R. W. F. Harrison, Assistant Secretary of the Society, 1896–1920.

to say if they are such as to make the following plan feasible: Harrison to be responsible for all the clerical work, the officers when they want work done which requires clerical assistance to give it to Harrison, tell him when they want it and leave it to him to arrange it in the way he likes best, provided it is done well and up to time. Do you think the staff is big enough to cope with all the work the war has brought? It seems to me that there is a danger of a breakdown from overworking some of the staff, if this took place it would make the satisfactory performance of the war work difficult. I think it is far more important for the Royal Society to do this work as well as possible than to save £200 or £300 in office expenses, and I think it would be wise to increase the staff for a time so as to be able to deal expeditiously with the war work. You will be able far better than I [to judge] if an increase is necessary, but if it is, I should think it bad policy to stop it because it costs a little money.

Excuse this rather incoherent scrawl but I am writing against time, one of our minor worries of the war is that they collect letters a good deal earlier than they did.

To Sir Arthur Schuster:

> *Forest House, Ashhurst, New Forest, Hants.*
> *Aug. 30th, 1916.*

Glazebrook has written a paper setting forth objections from the point of view of the N.P.L.* to the establishment of a physical laboratory for the Navy. I have written to him to the effect that if he sends it to any officials he must point out that it merely represents his own views, and has not been endorsed by either the executive committee of the N.P.L. or the Council of the Royal Society. I told him too that it would in my opinion be very detrimental to the interests of the N.P.L. to oppose on the ground of a problematical reduction in fees a proposal intended to benefit the Navy.

Like other presidents of the Society both before and after, Thomson had to try to satisfy malcontents, who had a feeling that power and influence was too much concentrated in the hands of the few. There has often been a tendency on the part of the scientific public generally to think that the fellows of the Royal Society have an undue share of influence. The fellows often think the same of the council, and the council think the same of the officers. The late Lord Oxford recorded that in his political experience he had

* National Physical Laboratory.

often thought that at last power was in his hands, but it always eluded him. This experience is probably general. Circumstances dictate within narrow limits what must be done and there is little scope for personal choice or the exercise of power. This applies least perhaps to the filling up of appointments, but here too, though the qualifications and limitations of individuals cannot be debated publicly before large bodies, in practice no one wishes to take the responsibility of making an appointment which provokes criticism or may prove a failure. That is the best security against the exercise of arbitrary power. The criticism was often made that members of the council had too short a tenure to get a real knowledge of the Society's affairs, or to become independent of the officers, but if a change were made, there can be little doubt that the opposite criticism would become vocal, and it would be complained that the council was a narrow clique to which the majority of fellows had no chance of admission. The truth is that the official business of the Society has not enough scope to exercise a great number of active and able minds and therefore if many such aspire to take an important part in it they are bound to be disappointed.

A correspondence arose in the course of the Royal Society business which is of some human interest. A change had been proposed by the council in the statute regulating the admission to the fellowship of a few distinguished men other than those possessing the normal scientific qualifications. This passed after submission to the body of the fellows in the ordinary way, no objection being made; but owing to war conditions some of the fellows did not realise in time what had been done, and when they did realise it, objected to it strongly. The question was reopened, and a special general meeting of the Society held to discuss it. In the course of the meeting Sir James Dewar made a speech which, like some other pronouncements of his, was not distinguished for lucidity. He referred to a document which had been quoted in the council's report and which he alleged had had its meaning altered by a change of punctuation. No one had been in a position to deny this at the moment, but afterwards it was found that there were in fact two versions of this punctuation, but that the council were relying on the original authentic version, and Dewar on the

incorrect one, which had no doubt arisen by inadvertence in re-
printing. Since J. J. Thomson and others (myself among them)
had understood him to suggest that the council had deliberately
'faked' the punctuation Thomson considered that the council
ought not to sit down under so preposterous an allegation, the
more so that they were correct and Dewar mistaken. One worldly-
wise member of the council urged strongly that it would be better
to pass it over in silent contempt, but the President and the
majority of the council thought otherwise, and in consequence
he was soon entangled in the morass of Dewar's irrelevancies like
a fly in treacle. The correspondence is too long to quote in full,
but the following passages give the gist of it.

THOMSON *to* DEWAR:

June 23, 1917.

I am instructed by the Council of the Royal Society to write you
on a matter arising out of the speech you made, at the Special General
Meeting of the Society on June 7th.

In that speech you accused those responsible for drawing up the
statement on Statute 12 which was circulated to the Fellows of the
Society, of altering the punctuation in the quotation, given on p. 2
of the statement, from the report of the 1875 Committee and thereby
misleading the fellows as to the view of that Committee.

There was no time during the meeting to refer to the report of the
Committee from which the quotation was taken. This is printed in the
Council Minutes for 1875, which are open to the inspection of any
Fellow of the Society. Since the meeting the report has again been
examined and *the punctuation is identical with that given in the statement.*

As you have based a charge of bad faith against a number of Fellows
of the Society on the alleged change by them in the punctuation, the
fact that there has been no such change creates a position which seems
to call for further action on your part.

DEWAR *to* THOMSON:

June 27th, 1917.

Before I answer . . . I am entitled to know what Statute or part of
a Statute or Order of the Society am I to understand the present
Council are authorising you as President to proceed upon in addressing
me in the terms you have thought proper to adopt.

The reply was a copy of a resolution by the council saying that they considered no statutory authority was required for their action, and regretting that Sir James Dewar had not withdrawn the charge.

DEWAR *to* THOMSON:

6th July, 1917.

Pray give me the names of any Fellows who are prepared to swear that I ever used such language when I was referring to the added word and the comma and semicolon, that is the punctuation question in the course of my argument.

THOMSON *to* DEWAR:

July 19th, 1917.

Am I to understand that you did not intend to accuse those who drew up the circular of an act of bad faith towards the rest of the Fellows? As no reporter was present, it is impossible to be sure of the precise form of words used in the debate; the important thing is the interpretation put on your speech by those present who heard it. I certainly came away with the conviction that you had accused the authors of the circular of deliberately altering the punctuation. You spoke of this being done very cleverly (I and several of the Fellows present remember these words) thus it must have been deliberate, and deliberately to alter a quotation is certainly an act of bad faith.

DEWAR *to* THOMSON:

July 21st, 1917.

I must respectfully point out that your letter of 19th July does not answer the distinctive question put forward in my letter of the 6th July. Clever syllogistic artistry—founded on an interpretation (that is motive)—is presented by giving an inferential proof of a heinous act. Such demonstration could not be accepted in any court of equity.

After a further exchange of letters which do not seem to add much of substance, Thomson wrote on July 30th proposing to close the correspondence.

DEWAR *to* THOMSON:

Aug. 4th, 1917.

I have no wish to continue this illusive correspondence, but I must emphatically protest against what you are now suggesting as explanatory of its collapse, viz. that 'there are no signs of our coming to an agreement'. You began the campaign of autocratic warfare on behalf of the Council by an attack on my good name, which might well have emanated from a Guild of Teutonic Kultur. This deadly mixed infection must be killed out of the atmosphere of the Royal Society, so that English science may remain pure and undefiled. This action of the council forced me into a defence of my rights, founded on legality and justice. It would be impossible under such circumstances to entertain the idea of 'agreement'. Again to end by being sorry that I did not take 'the initiative in the matter under discussion' borders on the hypocritical, which, for obvious reasons I need not discuss. The whole correspondence must now be submitted to the council, and I shall await an early reply.

It does not seem that any reply was given.

More serious than this absurd episode, in which so far as appears no one supported Dewar, was an attempt to prevent the re-election of Sir Arthur Schuster as secretary of the Royal Society at the anniversary meeting on November 30th, 1917. The ground of this was that he was of German origin—and in fact he had been born in Germany, though apart from a year or two at German universities his whole life had been passed in England. All his personal and family interests were in this country, and his energies had been devoted to the service of British science. However, in the eyes of a small minority of the fellows this counted for nothing against the fact that he had been born in Germany. H.E. Armstrong, the distinguished chemist whom we have already met, was the protagonist of this movement. An attempt was made to canvass the fellows, or at any rate those resident in London, but it does not seem to have met with much success.

Holmleigh, West Road, Cambridge.   Nov. 25th, 1917.

Dear Schuster,

I have not written to you about the Armstrong question before because I knew you would take a too unselfish view and would look at it too exclusively from the point of view of the Society. I have consulted a number of people and they agree with me that it would be best both in the interests of the Society and for your sake to have the matter out on Friday. As far as I can gather it is very nearly a case of *Athanasius contra Mundum*, entirely so if you count Dewar as either an angel or a devil as he is the only supporter of Armstrong that I can hear about. I sent Armstrong yesterday afternoon the following telegram.

'Should not rule discussion on qualifications of officers out of order on Friday provided it was based on a motion of which notice had been given in time to communicate to Fellows.'

This will oblige him to make a motion which can be put to the vote.

As I read the Statutes, the Anniversary Meeting is technically a Special General Meeting—it is included in the chapter under that heading and must I think be governed by the regulations in that chapter—one of which is that no business can be brought forward unless notice has been given.

I hope you won't take this contemptible business to heart—if you had heard the warm appreciations of yourself and work which have been expressed since this commenced you would feel that you possess to a very high degree the confidence and esteem of everyone deserving of respect in the Society.

Yours ever, J. J. THOMSON.

In reply to the telegram quoted in the above letter, Professor Armstrong wrote:

35 Granville Park, Lewisham.   Nov. 26th, 1917.

I see no reason why I should adopt the course you now seek to impose upon me. It was fully open to you on Thursday under the Statute to permit the discussion of matters outside the report; yet notwithstanding my insistent request you declined to allow me to deal with a question which you must know has too long been hanging over the Society, though no one has had the courage to raise it publicly. You declined also to give me an assurance that I might raise it on Election day.

I feel that I was entitled to greater consideration as a fellow of over 40 years standing.

To G. P. THOMSON:

Holmleigh.   Dec. 1*st*, 1917.

We have just passed through what might have been a troublesome affair at the Royal Society. Armstrong tried to engineer a movement to turn Schuster out of the Secretaryship. Fortunately he showed his hand a few days before the Anniversary Meeting, and steps were taken to circumvent it: we had a tremendous gathering on the day of the election (Nov. 30) but Armstrong's courage failed him at the last moment, and he did not turn up, and finally only 7 voted against Schuster. I had expected to have a lively time in the Chair with points of order raised every minute, but everything passed off absolutely quietly without any discussion.

It is satisfactory to be able to add that after Schuster's term of office as secretary had ended in the ordinary way, he was elected as Foreign Secretary to the Society without opposition on November 30th, 1920. No better proof could be wished that the unjust agitation against him had no ground in any real conviction that he was unworthy of the confidence which had been placed in him. Thomson would have liked to see him President of the Society, but in the nature of the case very few can occupy this position, and no opportunity offered.

To G. P. THOMSON: Trinity Lodge.   July 30th, 1918.

To-morrow at the Royal Society some of the hot-heads are bringing in a motion to expel all German Foreign Members. I expect we shall have quite a lively debate, as one of the Fellows is quite a passable imitation of Horatio Bottomley when he speaks on the subject.

The annual presidential addresses which Thomson gave on the 30th November each year dwelt, as was natural, on the war situation in its scientific aspect—the weaknesses in our organisation which had been revealed, and the measures necessary for dealing with them. He emphasised in particular the need for special research departments in the navy and army, the need for fostering key industries, the importance for manufacturing industry of industrial research, and the need that the industries should co-operate in supporting it, the desirability of introducing men old enough to have proved their scientific capacity into certain posts in the civil service, and the financial needs of the universities, particularly in the support of pure science. All these matters were dealt with by the governments of the war and post-war period, and, speaking generally, on the lines that Thomson advocated.

The period towards the end of the war was one of transition in the organisation of scientific research. Up till that time science had in the main been maintained by what may be called private and amateur effort. There were exceptions to this rule. Greenwich Observatory was a Government institution from the time of its foundation in 1675, but even here the idea of wholehearted Government support was not readily accepted. For example, when the astronomer applied for instruments, the application does not seem to have been considered reasonable. Later on came in succession the Geological Survey and the Royal School of Mines, the Royal College of Science, Kew Observatory and the National Physical Laboratory, and the Imperial College of Science. But upon the whole English science was built up by men who worked at their own expense or in private foundations such as the Royal Institution. In many cases, as e.g. Cavendish, Huggins, Joule, they worked quite privately in their own homes. In other cases, e.g. Newton, Stokes, they were university professors, but their universities did not do much if anything to finance their researches or even to provide a place where they could be conveniently conducted. As the nineteenth century went on this state of things changed, but only gradually. About 1870 the universities began to have proper laboratories instead of the attics and cellars which the scientific professors had been able to get possession of because no one else wanted them. The Cavendish Laboratory was a good example, and the intention of the donor had been to provide apparatus as well as the building. But there was no provision for current expenses in apparatus except the students' fees, and the treasury was the Professor's own pocket. In the earlier part of this narrative we have seen how difficult the position was. It was of no use to appeal to the University, because its funds were those derived from its ancient endowments, and were bespoken for other purposes and other studies. In the meantime the claims of science were becoming more insistent. It was no longer true that most pieces of apparatus that were likely to be wanted could be bought for some such sum as £10 or £20. Moreover, research students were inclined to raise their standards, and although they had perhaps no power to insist, yet in practice the position of a professor in daily contact with students who felt that they were being expected to make bricks without straw became increasingly difficult. I remember J.J. speaking to me in this sense shortly after the war. He was contrasting the research students of the day with those of the period circa 1900. 'They expect', he said, 'to be given a properly designed piece of apparatus which will *work*.' In the earlier days we had to make it work ourselves, as best we could, whatever its natural deficiencies might be.*

* Cf. p. 73.

It must of course be remembered that with the advance of science the problems to be tackled become continually more difficult, and it is necessary to be provided with modern weapons with which to attack them. These modern weapons tend to grow in cost and complexity. In 1918 the time when home-made apparatus of the sealing wax and string type of construction would meet many requirements was rapidly passing.

All these conditions made it necessary to appeal for Government aid. In many cases the exigencies of the war itself had compelled the Government *in extremis* to turn to the universities for assistance, for example, in dealing with the problems raised by the use of gas in war, and no doubt circumstances of this kind prepared the ground for pressing the national importance of research.

In 1916 Lord Crewe, who was then Lord President of the Council, received a deputation from an ephemeral body called the Conjoint Board of Scientific Societies which was headed by J.J. Thomson, as President of the Royal Society. The occasion was of some importance because the Government announced the establishment of the Million Fund to aid co-operative research to be set up by the various industries. J.J. delivered a speech which seems to have made a considerable impression, and the most significant part of it may be reproduced here.

I should, with your permission, like to say a few words on the importance [to our industries] of research in pure science, Sir Maurice Fitzmaurice will deal with Research in Engineering and Professor Baker with that in Industrial Chemistry.

I shall, this morning, look on research in pure science from a frankly utilitarian point of view, not because I think it is the only, or even the most important aspect of the question, but because it is the aspect which must, quite legitimately, appeal most forcibly to the Government responsible for the material welfare of the country.

By research in pure science I mean research made without any idea of application to industrial matters but solely with the view of extending our knowledge of the Laws of Nature. I will give just one example of the 'utility' of this kind of research, one that has been brought into great prominence by the War—I mean the use of X-rays in surgery. Now, not to speak of what is beyond money value, the saving of pain, or, it may be, of life to the wounded, and of bitter grief to those who loved them, the benefit which the State has derived from the restoration of so many to life and limb, able to render services which would otherwise have been lost, is almost incalculable. Now, how was this method discovered? It was not the result of a research in applied science starting to find an improved method of locating bullet wounds. This might have led to improved probes, but we cannot imagine it leading to the discovery of the X-rays. No, this method is due to an investigation in pure science, made with the object of discovering what is the nature of Electricity. The experiments which led to this discovery seemed to be as remote from 'humanistic interest'—to use a much misappropriated word—as anything that could well be imagined. The apparatus consisted of glass vessels from which the last drops of air had been sucked, and which emitted a weird greenish light when stimulated by formidable looking instruments called induction coils. Near by, perhaps, were great coils of wire and iron built up into electromagnets. I know well the impression it made on the average spectator, for I have been occupied in experiments of this kind nearly all my life, notwithstanding the advice, given in perfect good faith, by nonscientific visitors to the laboratory, to put that aside and spend my time on something useful.

This example illustrates the difference in the effects which may be produced by research in pure or applied science. A research on the lines of applied science would doubtless have led to improvement and development of the older methods—the research in pure science has given us an entirely new and much more powerful method. In fact, research in applied science leads to reforms, research in pure science leads to revolutions, and revolutions, whether political or industrial, are exceedingly profitable things if you are on the winning side.

Granting the importance of this pioneering research, how can it best be promoted? The method of direct endowment will not work, for if you pay a man a salary for doing research, he and you will want to have something to point to at the end of the year to show that the money has not been wasted. In promising work of the highest class, however, results do not come in this regular fashion, in fact years may pass without any tangible results being obtained, and the position of the paid worker would be very embarrassing and he would naturally take to work on a lower, or at any rate a different plane where he could be sure of getting year by year tangible results which would justify his salary. The position is this: You want this kind of research, but, if you pay a man to do it, it will drive him to research of a different kind. The only thing to do is to pay him for doing something else and give him enough leisure to do research for the love of it. Now this kind of research has been done in the past and will, I think, for some time to come continue to be done mainly in the Universities; and the best way to promote it would be to ensure that University teachers have leisure and opportunities for research and that their salaries are not so low that they have to spend all their spare time in examining if they are to earn enough to live upon. These men, however, require laboratories as well as leisure, and one of the results of this progress of Science has been to make laboratory equipment and apparatus far more costly than was the case a generation ago. Our Universities, as no one knows better than you, my Lord, suffer from an 'eternal lack of pence' and have not always at their command the funds necessary for the proper equipment of their laboratories. Money could be well spent in helping them to do this. There are other laboratories besides those at the Universities which are hampered by lack of funds. The National Physical Laboratory cannot afford to pay adequate salaries to its assistants, with the result that many highly trained men have left or are on the point of leaving to take more highly paid posts with private firms.

The amount required to assist research in pure science is but small in comparison with that required for Industrial research—the modesty of its requirements puts it in some danger of being overlooked altogether. (In farming the cost of the seed is not a very considerable item in the cost of the production of crops.) But pure science is the seed of applied science, and to neglect it would be on a par with the action of the farmer who spent large sums on ploughing and manuring his land and then omitted to sow the seed.

It was judged at the time that this speech strengthened the hands of the Advisory Council of the young Department of Scientific and Industrial Research (which was under the Lord President) in their policy of making grants to young and promising graduates in science to enable them to stay on at the University after graduation and to work under a professor who was active in research, the subject of the research being left to the student and his professor.

As regards the question of research departments in the Services, some wished that these should to a certain extent be under the scientific control of the Royal Society, as in the case of the National Physical Laboratory. Thomson seems to have been opposed to this, thinking that they were far more likely to succeed if the Services could regard them as part of their own organisation than if they had any element of alien control. No doubt this was the fruit of his experience at the B.I.R., and there can be little doubt that he was right.

On the other hand, he was not so clear about other Government research activities. In this connection there remains something to be said about Thomson's connection with the Department of Scientific and Industrial Research. This department was created to embody, unify and expand various Government scientific activities. Though technically it was under the Lord President of the Council, its powers were and are largely deputed to the Advisory Council, consisting of representatives of pure and applied science. Thomson was perhaps at first inclined to be suspicious of the Department, perhaps partly because some influential fellows of the Royal Society had hoped that research for Government might be developed through that Society, and partly because there was an idea that the new organisation would limit the freedom of universities. The National Physical Laboratory had grown too large a financial responsibility for the Royal Society to carry, and it was put under the Department so far at least as finance was concerned, though the Royal Society retained scientific control. In the transition, there was a good deal of friction between the Director and the officials of the Department. The Director (Glazebrook) found it hard to submit to Treasury control of finance, particularly in the more severe form which it assumed in peace time, when he had for many years enjoyed comparative independence under the Royal Society. In these circumstances the attitude taken up by Thomson made a good deal of difference, at any rate as to the entry of the new régime into smooth waters. He spent an afternoon with the papers by himself, and seems to have considered the official view, represented by Sir Frank Heath, the secretary of the Department, as justified. Indeed, it is believed that Glazebrook himself ultimately saw that there was no real alternative. Thomson indicated his attitude by hinting that he would be glad to become a member of the Advisory Council, to which he was promptly appointed. He threw himself into the work of the Council and rarely missed a meeting. The officials who worked with him have a lively recollection of his readiness to do all he could to make the work a success, and especially to maintain a high scientific standard. He was wont to emphasise that the best thing that science could do for industry was to introduce scientific *method*. His service on the Advisory Council was from 1919 to 1927.

In 1919, when scientific men were turning back to their own pursuits once more, Thomson dealt in his annual address to the Royal Society with Einstein's theory of relativity, which was then very much in the forefront, owing to the sensational results obtained at the eclipse of May 29th of that year, when it was shown that a beam of light was subject to gravity, and to just the extent that Einstein had predicted. Thomson accepted these results and the interpretation put upon them, but he never seemed particularly enthusiastic on the subject nor did he attempt to develop it, either theoretically or by experiment. I believe, from a conversation which I can recall, that he thought attention was being too much concentrated on it by ordinary scientific workers, with the neglect of other subjects to which they were more likely to be able to make a useful contribution. His attitude towards relativity was that of a looker-on. Probably the same was true of nearly all his contemporaries. It was the creation of a younger generation.

In particular, he did not take the fashionable view about *ether*. The following extract from a letter of 1937 to a correspondent* puts his view about this in a small compass, and is inserted here, rather out of place. Incidentally, his views about some other matters are included.

It was, I think, one of the great merits of Maxwell's work that he brought the subject [of electromagnetism] under the laws of general mechanics and not merely of the conservation of Energy which really cannot do very much in many cases; for example, though it could tell the speed of a body moving under an attractive force, it cannot tell the direction in which it would move. I take the view that a theory should be a policy and not a creed, that its most important work is to suggest things which can be tried by experiment, and for this the theory should be one that is easily visualised. It was Faraday's visualisa-

* Not the present writer.

tion of lines of force which led to the discovery of Electromagnetic induction. The mathematicians of the time were very contemptuous and asked why did one want more than the law of inverse squares which had been verified to I forget how many places of decimals.

Again, I differ from you about the value of the conception of an ether, the more I think of it the more I value it. I regard the ether as the working system of the universe. I think all mass momentum and energy are seated there and that its mass momentum and energy are constant, so that Newtonian mechanics apply. I regard the lines of force as linking up what we call matter with ether, that these lines like Vortex rings in air or water carry with them a volume of surrounding fluid much greater than their own volume, so that a part of the mass of the ether is linked up to the body from which the lines start and constitute its mass; this mass has, however, been taken from the ether so that the sum of the mass of matter and free ether is constant.

From LORD SANDERSON, G.C.B., Permanent Under-Secretary for Foreign Affairs, 1894–1906:

65 Wimpole Street.   April 7th, 1921.

My dear Master,

I avail myself of our old friendship at the Royal Institution to appeal to you for a small service, and hope that if you are too busy to attend to my request you will at all events forgive it. The circumstances are as follows:

Haldane* has been fascinated by Einstein and has retreated to the summit of the theory where he breathes the pure air of Relativity and looks down with compassion on his fellow creatures still floundering in the Serbonian bog of Newtonian and Galilean physics.

He is writing a book on Relativity as applied to ethics and politics, and has expounded its plan to Edmund Gosse, who, however, remains in a state of arrested development, owing, as he avers, to Haldane having fallen asleep in the midst of his discourse.

The Archbishop of Canterbury† on the other hand can make neither head nor tail of Einstein, and protests that the more he listens to Haldane, and the more newspaper articles he reads on the subject, the less he understands.

I am, or believe myself to be, in an intermediate stage, roaming about the lawns and meadow leazes half way down—

I therefore offered to write for the Archbishop a short sketch of what I imagined to be the pith of the theory in its most elementary form.

* The late Viscount Haldane, O.M., Lord Chancellor, 1912–15 and 1924–25.
† Archbishop Davidson.

I enclose it with his comment. It is of course very inadequate, but I fancy that as far as it goes it is not entirely at variance with Einstein's argument—some of his followers and critics seem to me to go further.

But I should be sorry to have misled the Archbishop. Do you think you could glance through it, or ask some expert to do so, and write a short note of any 'grosser errors'.

In return for this kindness I will with pleasure place at your disposal for my benefactor or for anyone to whom it would be of interest, a set of photographs of the Derby-Disraeli cabinet of 1866–9, which forms an agreeable companion to the last two volumes of Burke's Life [of Disraeli] . . ..

Sincerely yours, SANDERSON.

Thomson vacated the Presidency of the Royal Society on November 30th, 1920, at the end of the usual five years. It was agreed by the Council that he should write to approach Arthur Balfour with the proposal to become his successor. The latter, however, excused himself on the ground that he was already over-loaded with public work; moreover, he doubted the wisdom of appointing anyone who was not himself a scientific worker. Sir Charles Sherrington, O.M., the eminent physiologist, was eventually elected as Thomson's successor.

CHAPTER XII

APPOINTMENT AS MASTER OF TRINITY

EARLY in February 1918 Thomson was offered by the Prime Minister, Mr Lloyd George, the Mastership of Trinity College, vacant by the death of Dr H. M. Butler. The letter arrived one morning when he was just about to start for London. He ran upstairs to tell Lady Thomson. She naturally wanted to know whether he would take it, but he said he had not made up his mind. However her intuition was that he would accept and it proved to be right. Before he returned to Cambridge that evening he had already written from the Athenaeum Club accepting the offer.

It was naturally of interest for this book to find out what were the considerations which guided the Prime Minister's choice. In answer to a question Mr Lloyd George kindly wrote (Jan. 17th, 1941):

I have no recollection of having consulted anyone in particular about the appointment of Sir J. J. Thomson to the Mastership of Trinity College. His super-eminence as a scientist was known, even to a barbarian like myself who never had the advantage of any university training. As one of the War Directorate I knew what invaluable services Thomson had rendered in the conduct of the war. But I am sure that I would have talked it over with Lord Balfour, for whom I had the greatest respect as a counsellor and whom I met almost daily as a member of the War Cabinet of which I was the head.

At the time Thomson heard from a friend, probably well informed, at the Cloisters, Windsor Castle:

Arthur Balfour and the Bishop of Ely [Bp Chase] have been your very good friends.

In the same letter it was obscurely hinted that there had been attempts to push other claims to the Mastership.

The appointment is a very valuable one, in income amounting to about £3200, with the Lodge, rent, rates and repairs free. Moreover, at that time it was tenable for life. The Master of Trinity is better housed than the head of any other college at either of the ancient Universities, and has an excellent private garden, bordering the river Cam. Moreover, the historical tradition of the Mastership and the size and importance of the College make it a very dignified position. It is therefore not surprising that there were other aspirants, but they scarcely had a chance. Thomson had been closely connected with the College all his life, and he had contributed to the intellectual distinction of the College and the University in a way that went far to make his appointment a foregone conclusion.

The event brought him shoals of letters and telegrams of congratulation. Two may be selected for reproduction here.

House of Lords.   Feb. 13, 1918.

Dear Sir Joseph,

May I congratulate you on your accession to what has always seemed to me the most dignified of all academical posts not only in England but in Europe, for there is nothing on the continent that comes up to the headship of such a college as Trinity, with the splendid traditions that attach to it. Your predecessor was one of my oldest and best friends, and it is a pleasure to see his post so worthily filled as it will be by you. The world of scientific men will rejoice to see you where they would have wished you to be: and I trust that the duties you will have need not interrupt your work of scientific investigation and discovery.

Believe me,

Yours sincerely, BRYCE.

Ferne, Donhead, Salisbury.

Dear Sir Joseph,

I am as delighted as if it were myself Master of Trinity!

Don't trouble to reply.

Yours joyfully, FISHER.

10. 2. 18.

I am convalescing by leaps and bounds, and increasing my weight prodigiously.

The due acknowledgement of all the congratulatory letters, amounting to many hundreds, was no small undertaking. He did a dozen or so whenever he could snatch an odd half hour. They were all written quite personally to suit the recipients, and were not particularly short. Lady Thomson helped him, and incurred the penalty of writer's cramp from the amount of work to be done.

Superficially, Thomson was a great contrast to his predecessor, and seemed to belong to a widely different epoch. Butler was brought up in the classical tradition, and, according to stories repeated among his junior colleagues, was naïvely ignorant of scientific matters, though not on that account the less proud of the achievements of the College in the scientific field. Thomson was equally ignorant of the classics, though he too was not disposed to undervalue the achievements of which he knew nothing in detail. Butler's presence was large, slow and dignified in speech. Thomson was below the medium height, quick in speech and movement and not disposed to think much about dignity, and indeed not averse to raising a laugh by using an unconventional expression. Butler's conversation tended to edification, while Thomson's leaned rather towards gossip. It may be that these contrasts introduced a certain element of doubt when his appointment was announced, more perhaps on the classical and historical side, where Thomson's qualities were not so well known. But, after all, in many essentials the two men were comparable. Each had conducted a large institution successfully before he came to the Mastership. Butler could command the respect of those athletically inclined by his own youthful prowess, Thomson by his interest and knowledge of detail. Each was sincerely religious. Each was generous, hospitable and friendly, though each could be firm when the occasion needed it. Each was a ready speaker. Finally each was devoted to the interests of the College.

If there were doubts at first in any quarter the history of Thomson's Mastership was the story of his gradual establishment of a secure position in the respect and even affection of most of the Fellows, and of those many undergraduates with whom he made it his business to establish contact.

To G. P. THOMSON:

Feb. 26th, 1918.   *Cavendish Laboratory.*

I do not expect that the Mastership will diminish the time that I can give to science. I am determined that it shall not, and I think it will turn out that it gives me more rather than less. The college has been very kind and they say the last thing they would wish would be for me to slacken off scientific work. The actual work the Master has to do does not demand a great deal of time, and, what is much to the point, is done in college so that I have not as in some of my work in London to spend more time in getting to and from the work than over the work itself. I have too a great deal less work in London than I had—the Education Committee has finished, and we have reorganised the work of the Board of Invention so that the number of our meetings is much reduced. My chief work in London now is the Royal Society. With regard to the Professorship I am going to resign the salary, and ask to be relieved of the courses of lectures for the Tripos. I hope, however, to retain the control of the Laboratory and research work; this will be all the change that I think will be necessary during the war, afterwards for a permanent solution something more will have to be done. I think myself there ought to be two Professors of Physics.

I wish you could have been here to the admission which takes place next Tuesday, there is a most elaborate ritual connected with it, it takes up about 6 type written pages, and the leading principle seems to be that in all the processions and evolutions which precede my making the declaration I must manage to dodge so as to be on the left hand of the Vice-Master while after the declaration I have always to be found on his right.

The following detailed account of the installation ceremonies is derived in the main from *Cambridge Notes* by W. W. Rouse Ball (Cambridge, 1921), though a few sentences have been interpolated:

On the appointed day all entrances to the College were closed and locked at 10.30. At 11.30 the great bell was rung, and continued to be rung, until the master-designate entered the College at about 12.10. Members of the College, visitors who had obtained orders of admission, and a considerable number of army cadets then stationed in it began to assemble in the great court soon after 11.0, and by request kept to the south of the pathway from the great gate to the Master's lodge, and to the west of that from the sun-dial to the clock tower, these forming the processional route. At 11.45 the fellows in their proper academic dress assembled in the ante-chapel—the vice-master and deans wearing in addition their hoods but not squared. The family of the master-designate occupied the room above the great gate.

At noon, Sir Joseph Thomson, wearing cap, gown, bands and his hood squared, came to the double doors of the great gate and thereon knocked loudly three times. The head porter, Coe, wearing his robe

J.J. Thomson waiting at Trinity Great Gate, previous to his admission to the Mastership, 1918

of office, opened one door, and, holding it ajar, said, 'Who are you, Sir?' As Coe had for years been gyp in his college rooms, the question was slightly superfluous. However, the master-designate replied, 'I am the Master of Trinity'. Coe then asked, 'Have you letters patent from the Crown appointing you?' Thomson answered, 'Yes', and thereon handed to Coe a large unsealed envelope containing the letters. Coe received them, and said, 'I will take them, Sir, to the vice-master and fellows'; the gate was then closed, leaving the master-designate outside, the object of much critical observation. A considerable force of police was on duty and kept for him a free space during his vigil. W. H. Hayles, the lecture assistant at the Cavendish Laboratory, took a photograph by permission.

To return to affairs inside the College. After receiving the letters patent, the head porter placed them on a silver salver and, carrying his mace, proceeded by the processional route to the ante-chapel, where he handed the letters to the vice-master. It is interesting to note that the long staff bearing the college arms, used generally by the head porter on formal occasions, is, on the admission of a new master, discarded for a short mace about a foot long, topped by a gilt crown, the whole being something like a marshal's baton: this mace is used on no other occasion.

In the ante-chapel the vice-master received the letters, and, having previously had an opportunity of examining them at leisure, briefly stated that they granted the vacant mastership to Sir Joseph Thomson, were sealed with the great seal, and apparently were in due form. The Society then proceeded to the great gate to receive and welcome the master-designate. The head porter walked first, followed immediately by the vice-master, carrying the letters patent in one hand, and with the senior resident fellow on his left, next came the deans, and after them the other fellows, two and two, in order of seniority.

On reaching the great gate the fellows took up position on the south of the processional route, and the vice-master, moving forward alone, said to the head porter, 'Porter, open the great gate'. On this being done, the master-designate entered, and the vice-master, removing his cap, said, 'Master, we welcome you. Master, I have the honour of presenting you to the fellows of the College', on which general bowing ensued. The procession then reformed, the senior fellow falling back to walk with the deans, and the master-designate walking on the left of the vice-master: in this order, and preceded as far as the chapel porch by the head porter, it crossed the court. The fellows then entered the chapel, and occupied their stalls in order of seniority, no one else being admitted. As soon as the master-designate had entered the College, the gates were again closed so as to prevent the outside crowd invading the courts.

After the fellows had taken their positions in chapel, the junior bursar locked the inner doors under the organ screen, and the chapel clerk the outer doors in the porch. What then took place is not officially known. The *Cambridge Chronicle*, however, issued the next day an inspired statement that 'the vice-master read the patent to the assembled fellows, standing in their stalls, and the new master made the declaration required by the statute and signed it, his signature being witnessed by the vice-master. The junior bursar then opened the doors of the master's and vice-master's stalls, and the vice-master, taking the master by the right hand, conducted him to his stall and admitted him to the mastership in the prescribed Latin form', which, we may conjecture, ran in the words, *Auctoritate mihi commissa admitto te in perpetuum Magistrum Collegii Sanctae et Individuae Trinitatis Cantabrigiae*.

At the end of the installation, the junior bursar opened the inner chapel doors, the organist went to the loft and began playing, the outer chapel doors were opened, and at the same instant the college flag (France ancient and England quarterly) on the great gate was broken. The chaplain (only one being in residence), librarian, and choir, who had assembled in the vestry, then entered the chapel followed by members of the general public who were in the court and cared to come. Next the *Te Deum* was sung. After this the doors of the stalls belonging to the master and the vice-master were opened, and they, followed by the fellows, descended and formed again in procession as before—the new master of course taking the head, and placing the vice-master on his left. The chaplain, librarian, and choir followed the fellows out of the chapel, but turned into the vestry. At the chapel porch the head porter met the procession and preceded it by the sun-dial to the porch of the master's lodge, where it broke up. Here for a few minutes the new master stood, receiving the congratulations of personal friends; after which he entered his lodge, he himself opening the door, thus terminating the ceremony.

J.J.'s speech at the dinner in the Hall that evening was thought to have been one of the best he ever made. Unfortunately it was not recorded. He spoke of what he considered the setbacks of his earlier life, such as the failure to get more than a minor scholarship, and his failure to get the professorship at Manchester. He also spoke of the Fellows who were in residence at the time when he came up himself.

Sir George Trevelyan (père) wrote (March 8th, 1918):

I suppose I took my degree earlier than any other existing Fellow of Trinity. That circumstance explains my being unable to come, and enhances the vexation of being unable to testify in person to my pride and satisfaction in the good fortune of our college at securing you as Master. The pleasure to me is enhanced by a circumstance to which I should not allude, except privately and confidentially to anyone except a Trinity man *and an O.M.* It was a great satisfaction to dear Dr Butler that of the twelve or thirteen civil members of the Order, five were Trinity men: and I shared his satisfaction to the full when the list of the order was graced and strengthened by your name. I hope I may venture to say one word about the keen and even jealous interest with which I awaited the choice of a successor to my dear friend....On last Christmas day I wrote to him after reading with intense interest Monk's life of Bentley with Macaulay's [manuscript] marginal notes, looking forward often to see how many pages more were left of a book which could not be too long for a Trinity man....Amongst other topics in his letter was a good deal about the North staircase in the Lodge, which was the first cause of the forty years war between Dr Bentley and his Fellows. Coming fresh from the strangely fascinating book I appreciate all the more the importance of an office, which from a boy upwards, my uncle* taught me to regard with no common feelings.

Thomson's eventual successor, G. M. Trevelyan, wrote from Italy (March 10th, 1918):

Apart from the desire to show respect to the college and to yourself, I cannot conceive a ceremony I would more have wished to witness with its moving contrast between the peaceful pageantry of the old academic world, and the tragic realities of the situation, collegiate as well as national, in the period when you are called upon to preside over the fortunes of the college. That we have so distinguished a head will be of good augury for the reconstruction period.

Punch, February 20th, 1918. From a note on the new Master of Trinity:

'Among his many scientific achievements was the discovery of the nature of the catholic rays which are generated by electric discharge through a vacuum.' Morning Paper. Surely the last word must be a misprint for 'Vatican'.

* Lord Macaulay.

To G. P. THOMSON:

> *Trinity Lodge.   May* 19*th,* 1918.

We are now practically settled in the Lodge, it is a most delightful house, and we are continually discovering fresh things of interest. The judges come here next week, and we have to abandon the lower part of the house to them for the two or three days they are here. I have all kinds of odd jobs to do as Master, I have just come from chapel, where I had to read one of the lessons as it was Whit Sunday. The work of Master is very interesting, and does not take much time—any meetings there are are held either in the combination room or the Lodge, so no time is wasted in getting to them. I must now rush off to read grace in Hall.

Holmleigh was let for a time to a Government department in 1918, the Thomsons retaining the garden to grow fruit and vegetables. Thomson had been at first averse to selling it, and indeed no one can very much like to part irrevocably with what has been a happy home. However, the lease was sold a few years later.

This will be the place to say something about the Master's Lodge, which was to be Thomson's headquarters for the rest of his life, as the Cavendish Laboratory had been for the earlier half of it. The Lodge faces the visitor across the great court as he enters the Great Gate. The large entrance hall contains a plaster cast of the effigy of Francis Bacon in the ante-Chapel. It is panelled in black oak, as is also the main staircase. The latter is hung with an extensive collection of photographs and engravings of famous Trinity men, collected, I believe, by Dr Butler, who was accustomed to entertain his guests with interesting reminiscences of the heroes they represented, most of whom had been known to him. I do not think J.J. ever followed Dr Butler's tradition in this matter, though he made a few additions to the collection. On the landing is a fine pendulum clock, presented by Sir Isaac Newton to the observatory over the Great Gate and transferred to the Master's Lodge when the observatory was dismantled. The dining room, like the entrance hall, is panelled with varnished oak, and adorned with pictures of the various masters of Trinity from Elizabethan times onwards. Through the dining room is a passage leading to the judge's bedrooms, formerly used by two judges, but now by the judge and his marshal. The upstairs drawing rooms are two; the second and larger, over the dining room, is a magnificent Elizabethan room with an original ceiling which, it is said, was long concealed by a plain ceiling below. All memory of it had been lost, until it was accidentally rediscovered. This room too is hung with portraits of Trinity celebrities, and Royal patrons.

For those who wish for details, the history of the Lodge as a structure is given in *The Architectural History of Cambridge* by Willis and Clark (Cambridge, 1886), which takes us down practically to the condition of affairs when the Thomsons came in—for Dr Butler had not made any important changes. The story is very much involved with the war between Bentley and the fellows, referred to in Sir George Trevelyan's letter above quoted: for their chief grievance against him was that he encroached on their amenities in order to improve his own, and improperly expended College money in the improvement of the Lodge.

In Butler's time and in Thomson's, the current was perhaps setting a little in the opposite direction. The College wished to make a new combination room and parlour at the expense of the Lodge. It could only be done by the Master's consent, and the proposal was mooted before the war, but Dr Butler begged that in view of his advanced age it might not be pressed. In 1920, when the war was over, the scheme cropped up again, and Thomson was approached. His attitude may be described as one of toleration towards it; in fact he did not care to refuse what was generally wished. He personally liked the old combination rooms, large and small, on the south side of the hall, i.e. on the side remote from the Lodge, and regretted their being abandoned. There was, however, some advantage for the fellows in having larger rooms, and it was an undoubted convenience to have them at the same end of the hall as the high table, without the length of the hall and the screens intervening. To gain the necessary space, some rooms in the Lodge had to be sacrificed. On the ground floor the kitchen and the servants' hall of the Lodge gave place to the fellows' parlour, and on the first floor, the old master's bedroom and study gave place to the new combination room. Space was found for the new kitchen and servants' hall by remodelling the old servants' offices on a smaller scale, the new kitchen being very

small compared with the old one. Further, the domestic arrangements were very much behind the times. Hot water could scarcely
be had except on the ground floor, and there was very poor
central heating. These deficiencies were made good.

The actual building operations were necessarily a source of considerable annoyance to the inmates of the Lodge, and the College,
in order to spare them this, wished to make provision elsewhere.
Mr W. W. Rouse Ball, a former tutor of the College, and a well-
known writer on subjects connected with mathematics, generously
offered his house for the purpose, but Lady Thomson wished to
be on the spot to supervise alterations, and the Master had the
strongest objection to being uprooted; so they stayed in the Lodge,
and bore with the noise and discomfort as best they might.

In course of the alterations a curious discovery was made in
the wall between the small drawing room and the old study—the
same wall which now separates the combination room and the
small drawing room of the Lodge. The panels behind the fixed
bookcases were found to be hollow, and the hollow space re-
sembled a sentry box in shape. It was built in with panels all
round, except for a round hole about one-third of the way up
from the ground. No satisfactory explanation has been suggested,
though the space is reminiscent of a 'priest's hole'. Such hiding
places were provided in troublous times as a general insurance,
and this (if it was one) need have had no particular relation to the
political or religious views of the Master and Fellows of the day.

It was necessary to provide a substitute for the old master's
study which had been sacrificed, and this was done by taking
in and converting a set of fellow's rooms on the other side of
the Lodge, round the corner of the great court. In this way
a spacious room was provided, with access to the great court up
a separate staircase. The door on to this staircase was usually kept
locked, but on occasion Thomson found it convenient to get in
that way if Lady Thomson had a meeting in the drawing room,
or if there were visitors who he thought might detain him longer
than he wished. Members of the College often used this door
when they came to see him on business. The new study was lined
with oak bookshelves.

CHAPTER XIII

THE EARLIER POST-WAR YEARS

As we have seen (p. 208) J.J. had not contemplated holding the Cavendish professorship beyond the end of the war. In March 1919 we find that he had resigned it, and that the University had appointed him Professor of Physics (unpaid). He wished to continue his experimental work at the Cavendish Laboratory and to take some share in directing research work there, though he doubtless felt that to direct thirty or forty research students, as he had done before the war, was too much when added to the duties of the Mastership. The succession to the Cavendish professorship was obviously open to Rutherford if he was willing to take it, but to delimit the exact provinces of the former Cavendish Professor and the new in directing the work of the laboratory was a delicate matter, and it is evident from the letters which follow that Rutherford felt this, and that Thomson was most anxious to satisfy him.

Saville Club, 107 *Piccadilly, W.*

My dear Professor,

I have been thinking over the Cavendish matter but of course there are a number of factors that enter into the question. Before coming to any decision there are several important points on which I would like your views and frank opinion.

Suppose I stood and were elected, I feel that no advantages of the post could possibly compensate for any future disturbance of our long continued friendship or for any possible friction, whether open or latent, that might possibly arise if we did not have a clear mutual understanding with regard to the laboratory and research arrangements. It is for these reasons that I feel it very desirable to discuss the position in its various bearings. In the first place, I should say that if elected I would welcome your presence in the Laboratory and would be only too happy if you would help us as far as you feel able in helping research and researchers in the Cavendish.

I feel confident that in the near future there will be more advanced students than the Lab. can provide for, or that [*sic*] two of us could look after properly. Under such conditions it might prove awkward for both of us to place intending researchers in the position of making a decision with whom they wished to work. To avoid such undesirable complications there appear to me two possible solutions; either for you to work 'solus' with your laboratory assistants or for the Director to have charge of all advanced students and to assign their line of work and their supervisor. Under the latter arrangement, one would naturally place under your supervision those students who worked along lines in which you were specially interested or on topics in which you wished further investigation. If we kept closely in touch on projected lines of research, I would know your wishes and would try to fill them by turning over, as occasion arose, men to work under your supervision, but of course such students would be students of the laboratory first, and of their supervisors second.

How does the above appeal to you as a good working arrangement for both sides? I should welcome any other suggestions to meet possible difficulties.

Another point; the new director might feel it desirable and necessary to make changes of the organisation of the teaching and research in the lab. and possibly also in the personnel—with which changes you might possibly not altogether concur. It would be a disaster if any trouble should arise on that score, for it seems to me that the Director must take sole personal responsibility for the efficiency of the teaching and research in the lab.

I have spoken quite frankly of possible sources of misunderstanding, and I feel if I were elected it would be of the greatest importance to both of us to have such a clear mutual understanding of the situation and its inherent difficulties that we could work in complete harmony.

Another financial point; what is the income of the lab. and what payments are included in it? Does it have to cover not only for workshop, upkeep, and apparatus, but also for salaries of teachers? A rough idea of the situation in this respect would be very helpful.

I am afraid I am troubling you a good deal on a number of delicate points, but I would be very pleased to hear your views as soon as possible as I will have to make a decision whether I will become a candidate during the next few days. . . .

Yours ever, E. RUTHERFORD.

The reply to this letter is not to hand.

(No address, but doubtless from St John's College, Cambridge.)
March 23, 1919.

Dear Thomson,

As you know, Rutherford has been thinking seriously about the Cavendish chair. And there seems to be a strong feeling in the University that he should be attracted if possible, on the ground that your activities will be withdrawn more and more to other duties, and in any case can only last a limited time, after which the repute of the school now one of the essential assets of the University might fail.

I understand that he was in correspondence with you about the mutual independence of the two chairs and free scope for each. And he was quite satisfied, he said, with the result. But the subject has now recrudesced. I feel certain that it is being used as a lever by the people at Manchester to prevent him going on with the matter, in conjunction with assurances of complete facilities there. They have of course the object lesson of their two chairs of chemistry to work upon.

He puts it now that there is not room in the laboratory for two departments, to which I answer that it ought to be extended very soon; that there are not funds to equip two departments, to which I answer that the University is approaching the government; that he is an independent power and influence in Manchester, to which I am not in a position to make any answer.

The thing must settle itself in a few days. It is outside my province except that I happen to be an elector and the only local one whom he knows; and therefore I am bound to pull the threads together as far as possible.

This I can best do by telling you the facts and leaving it to your judgment whether you can do anything further upon them by way of communicating with him. He will not know that I have written to you unless you choose to tell him.

I am sorry to appear meddlesome, but it seems to be up against me.

Very truly yours, J. LARMOR.

March 23rd, 1919.

Dear Rutherford,

I am anxious to make the position with regard to the Cavendish Laboratory as clear as possible, even at the risk of repeating what I said before. The intention is to make the two Professors as independent as if their laboratories were in separate buildings and as soon as the necessary funds can be got the University will, I am sure, take steps to obtain a new laboratory. In the meantime a few rooms in the new wing of the present laboratory are assigned for the use of the new professorship, but these and these only will be subject to his control. Speaking for myself, I should never dream of interfering in any way with the rest of the laboratory or expressing an opinion about matters of policy. I should treat it just as I would a separate laboratory a mile away. The details as to the rooms to be assigned are not yet settled.

This awaits the appointment of the new Professor, and I shall be most anxious to fall in with his views. I regard the new Professorship which the University has asked me to take as being a detached and separate post and having, if I may so express it, no continuity with the Cavendish Professorship which I have resigned.

When I sent in my resignation I had thought out the question and determined to sever myself entirely from any connection with the management and policy of the Cavendish laboratory, so that my successor might have a perfectly free hand to carry out any policy he might see fit to adopt. I shall confine myself to my own rooms and do all I can to get a new Lab. built. I said most of this in my former letter, but I think it cannot be too often repeated that as far as I am concerned the new Cavendish Professor will be as independent as if he were in an entirely separate building.

J. J. T.*

I am sure you can rely on everyone in the University doing everything in his power to meet your view. There is a very keen hope that you may see your way to come to Cambridge.

To another possible candidate for the post, J.J. wrote from Trinity Lodge, March 28th, 1919:

I saw Rutherford yesterday, he had not made up his mind, and I honestly do not know what his decision may be. I am very anxious that the chair should have a worthy occupant and I am somewhat afraid that if the big men do not give some indication of their views before the election someone may be elected who is not the best we could get. There are great opportunities for making a very great school of Physics at Cambridge as we have for example young Bragg, Charles Darwin, G. I. Taylor, Aston, and others all in residence, and the University will I think do all it can for the subject.

No one can fail to be struck by the modesty of this letter. The author of it does not give the slightest hint of what was obviously the fact that he himself *had* made a very great school of physics at Cambridge, and that the problem for his successor was to maintain it.

In the event, Rutherford applied by telegram sent on the day of election. He was elected, and under him the laboratory main-

* This informal signature is from the rough draft. The copy actually sent is not in my hands.   R.

tained the brilliant reputation it had enjoyed under his predecessor. It will be seen that every attempt had been made to clarify the provinces of the two professors beforehand, and it need hardly be said that both parties loyally adhered to the arrangements agreed upon. There were, however, residual difficulties inherent in the situation. No arrangement as to who was to direct individual research students could in practice overrule their own choice. They had to decide for themselves. J.J.'s great achievements dated from the time when they were children, and he was approaching what is now regarded as the age of retirement for Professors. Under these circumstances it was inevitable that most of the students entering on research work should elect to worship the rising sun—and the rising sun could not well help it.

Before the war, Thomson had taken in the chief German publications on physics, and had systematically studied them. This kept him *au fait* of a constant stream of thought largely independent of the parallel development in this country, but very fertile in combination with it. During the war, these publications could not easily be procured, and moreover the value of their contents was for the time considerably diminished. Then, when the war was over, he was no longer mainly responsible for setting young graduates to work on research at the Cavendish Laboratory, so that there was not the same pressure upon him to keep abreast of current developments in order to find subjects for these. Thus it came about that he never took in the German periodicals again, and never resumed the habit of studying them systematically. Then again his duties as Master of Trinity, as well as other public duties, made competing claims upon him. These various circumstances led to his career as an experimentalist passing into a phase of diminished activity. Such experimental work as he did was mostly on the old lines, and was not marked by any distinctive development of technique. He had given his chief message to the world, and it did not suit his character or turn of mind to be anything but a leader. Leadership was passing into the hands of younger men.

J.J. became president of the newly created Institute of Physics which was organised to safeguard the professional interests of physicists. A passage from his address of 1922 will be quoted, which illustrates some of the difficulties with which scientific workers had to struggle in those days.

I wish to say just a word on the effect of the Safeguarding of Industries Act on research. It would be very improper for me from this place to lay down any political principles or to criticise the Act from a political point of view. I speak merely from my own point of view. I myself think that this Act has greatly increased the difficulties in this country. I am only giving my own experience. I have lost more time since the war by the use of imperfect materials than I did in the 40 years I have been working. Over and over again glass apparatus which has taken a fortnight or three weeks to construct has cracked during the night, and the whole process has had to be repeated.

A great deal is said now about key industries. Research has some claim to be considered a key industry, and it is entitled to the encouragement such as it is under the Safeguarding of Industries Act. It may be right to protect these industries, but this is a national question, and if the nation gains, every member of the nation should pay his share in any of the penalties which have to be paid. At present the burden falls on only a small section of the community instead of being distributed over the whole. It would not outrage any political principles if licences were given to recognised Institutions to enable them to import materials or instruments required for research. That, I think, would be a considerable improvement on the present position. For example, while galvanometers generally can be obtained, Paschen galvanometers can only be obtained from one place, and it seems a little hard not to be able to obtain one of these instruments without paying a heavy price.

In the course of time, these troubles were lightened by the development of British manufactures, and by the easing of the import restrictions when they press too hardly, so that later the matter became one of historical interest only.

At the time when Thomson became Master of Trinity in 1918, the question naturally arose of whether he would continue as Professor at the Royal Institution. The Managers of the Institution wrote pressing him to do so, and he agreed to carry on for the time, feeling perhaps that it would hardly be fair to them to compel them to find a successor in war-time. In 1920 this difficulty no longer existed, and he resigned the position, remaining, however, honorary Professor of Natural Philosophy. He spoke at the unveiling of the Memorial to Dewar there some years later, emphasising the value of Dewar's method of exhaustion by cooled charcoal, which he thought had been almost a main contributory cause to the progress of modern physics.

Towards the end of 1922 he was invited to give a course of lectures at the Franklin Institute of Philadelphia. He found it difficult to decide whether to accept; and at any rate the order for bulbs for the garden must be sent off first. The terms offered were very generous, including a large fee and travelling from Cambridge as guests of the Institute both for himself and his wife and daughter. Lady Thomson was unable to go, but Miss Thomson urged that she must come, as he would lose a pair or so of pyjamas everywhere he stayed if there was no one to see after him, and so it was arranged. He cabled to accept. They sailed towards the end of March 1923 in the *Majestic*.

He closely cross-examined some American fellow travellers who sat at the same table as to what kind of shellfish and other delicacies would be in season when they arrived. On the morning of the Boat Race the waiter brought J.J. some Oxford marmalade, but when he saw what it was he indignantly rejected it, and the waiter had to bring some of a different kind. In the meanwhile, Principal Irvine* (an Oxford man) and the American fellow traveller, who had been at Harvard, which that year supplied Oxford with their stroke, pounced upon the despised pot and finished it between them.

At New York men connected with the Franklin Institute met them, and J.J. plunged into conversation with them, leaving Miss Thomson to deal with the customs, who regarded the apparatus for his lectures with considerable suspicion. They were entertained *en Prince* for three days, which were chiefly spent by J.J. in seeing laboratories, and visiting Long Island. From New York they went on to Baltimore to visit Professor and Mrs H. F. Reid. Newspaper reporters were waiting on the doorstep, but they were put off until after dinner, when they were interviewed. Ten minutes after they had gone, the telephone bell rang, and one

* Now Sir James Irvine, Principal of the University of St Andrews.

of them said he had forgotten to ask a most important question. He had heard that J.J. was one of the few men in England privileged to keep his hat on in the King's presence. Was this true?

Another encounter with a newspaper reporter was at his rooms in the hotel at Philadelphia. J.J., pulling out a key from his pocket, found that it belonged to the Plaza at New York. This incident was judged of enough importance to afford material for a whole column of newspaper print.

J.J. went from Baltimore to visit Harvard, Yale, Princeton, and the Research Laboratories of the Western Electric and General Electric Companies, before going to Philadelphia for his lectures. These latter were given on five consecutive days in April. The subject was 'The Electron in Chemistry' and the lectures were subsequently published under this title by the Franklin Institute. In the preface he speaks of the experience of giving them as one of the most delightful in his life, on account of the boundless hospitality and admirable arrangements made by the members of the Institute.

In these lectures he proposes a model of the atom in which stability of the electrons is obtained by a different method from that of the sphere of uniformly diffused positive electrification which he had employed earlier. He had probably never been very well satisfied with this hypothesis, which really had no justification, except that it got over the mathematical difficulty of obtaining stability under the law of inverse squares. In the meantime Rutherford and Bohr had shown the advantage from quite another side of introducing the idea of a concentrated positive nucleus and the conceptions of the quantum theory into atomic models. J.J., though he was no doubt, like everyone else, well aware of these contributions, preferred to go on on lines of his own, and introduced a law of force which was such that the attraction of the nucleus for an electron changed to a repulsion at a distance of about 10^{-8} cm. On this basis he developed the theory of various atoms and molecules, and attempted to trace the relation between structure and various chemical properties. However, the basic hypothesis employed seemed to many to be too artificial to carry conviction, and, so far as I have been able to learn, the

J.J. Thomson with Dr Irving Langmuir in the Laboratory of
the General Electric Company, Schenectady, U.S.A., 1922

views suggested have not had any great influence on the subsequent development of chemical thought; it does not therefore seem desirable to attempt to explain them in detail here. The fact is that static models of this kind are no longer in fashion. It is generally felt that the mysteries of atomic constitution lie much deeper.

They returned to England on the *Homeric*, loaded with presents of flowers and boxes of chocolate. J.J. was more successful in defending these latter against the criticism of the customs officials than he had been in dealing with his own scientific luggage on entering the United States.

His impressions of the various industrial laboratories in America were summarised in an address to the Institute of Physics on his return: it seems better to quote the greater part of this address, which is interesting and characteristic, rather than attempt to summarise it.

I might perhaps say a word about the history of the industrial research institutions. On the present scale they are quite modern, even for America. They were not started without great opposition. The credit of starting them is, I believe, due to Mr Whitney and Mr Coffin of the General Electric Company who had to fight a long and strenuous battle with the shareholders and many of their co-directors, and for some time after they had established a research laboratory, they were, at the annual meeting when the accounts were submitted, exposed to the criticisms of their opponents, who wanted an immediate return for the outlay. Criticism was, I believe, especially directed against the nature of the work done by Mr X. as having no possibility of practical application; by the irony of fate this work developed into one of the most lucrative of the activities of the company.

I had the opportunity of talking with men, business men first and foremost, with no special interest in science, who were connected with the companies. They one and all said that the research department was one of the most profitable, if not the most profitable, department of their business, and would be the last they would reduce if expenses had to be cut down.

The scale of these laboratories is enormous. Take the example of the Western Electric Company, a branch of the Western Telegraph Company, which has an immense building for research and development. There is a staff of 3000 and the annual expenditure on this building is over £1,000,000 sterling. I spent a day going over the laboratories, and saw work going on a great deal of which was extremely interesting from the purely scientific point of view.

For example, there was a very interesting research on the effect on the perception of speech of cutting out by filters blocks of sound between different frequencies. A passage from a book was read and first one block of frequencies was cut out and then another. It was surprising how much could be cut out without altering materially the intelligibility of the speech.

Then, too, I saw in the laboratory interesting experiments of scattering of electrons by elements of different valencies.

I saw a new alloy which they invented, which had special properties. Under low magnetic force its permeability is enormous, greater than any known. It is of great industrial importance.

Then, too, they had a most ingenious way of uniting copper tubes by cementing them directly to glass without using wax joints, which are the source of so much trouble in work on gases at low pressures. I saw, too, enormous developments in wireless. I had the pleasure of speaking directly with San Francisco and Havana and listening in to a conversation between these two towns.

I spent two interesting days at the General Electric Company's laboratories at Schenectady. As far as buildings go, they are smaller than those of the Western Electric Company, but the Western Electric Company's work is partly development work while that of the General Electric Company is pure research. The staff is very large, I do not know the number but it was sufficient to fill a very large room, and the annual expenditure is something between £200,000 and £300,000 sterling. I talked to various members of the scientific staff who were working there and saw experiments they were doing. These in many cases were as purely scientific in type as those which go on in the Cavendish Laboratory.

They do not find any difficulty in finding practical applications for their discoveries. I found the staff enthusiastic and with a good knowledge of all the latest developments in their subjects and thoroughly interested in testing theoretical views; in fact, the general tone of the place is similar to that in the universities in England.

These research laboratories are attached to a single firm. This has a good many advantages, it gets rid of rivalries and jealousies between the various firms. The results they get are for the benefit of the firm providing the laboratory. The maintenance and equipment of a large research laboratory is, however, an expensive business and is beyond the powers of companies with only moderate capital. In England where enormously wealthy corporations are not so numerous as in America we must rely in the main on research laboratories which are supported by the industry as a whole and not by individual firms.

In my visit I was impressed not only by the commercial importance of these research laboratories but also with their scientific possibilities. In these laboratories things are done on an engineering scale and they have at their command currents, voltages, magnetic forces far larger than those available in university laboratories; they are able to get quite large effects when in an ordinary laboratory it would take a long time to make sure that the effect existed at all.

At the General Electric Company I suggested that it would be a great boon for science if these powerful appliances could now and then be available for a physicist who required for the test of some theory instruments more powerful than those at his disposal. They sympathised with that suggestion and said they would welcome workers of the type I had in mind.

I need not say much about the laboratories of the universities except that since my last visit about thirteen years ago nearly every university seems to have received from some generous donor a new physical laboratory. These are all well designed and very large, indeed I am inclined to think that some are too large and that it would have been better to spend a smaller amount on building and retain funds for buying apparatus for research and for the general development of the laboratory.

One reason why so many laboratories have been given to the Universities no doubt is that there are many rich men in America; another and, I think, perhaps even a stronger one is the feeling in regard to money which exists in that country. I do not know any country in the world where less deference is paid to mere wealth. In America, wealth can in some ways do very little; it cannot buy social distinction. This I am told is not quite true of New York, I have had very little experience of social life in New York, but it is certainly true in Philadelphia, Baltimore, and Boston, where I have spent a considerable time and know a good many people. So that if a rich man wishes to gain the esteem of his fellow-countrymen he does it by devoting some of his money to public use. The anxiety of many of their rich men seems to be to find some outlet for their money. For example: a man came to me and said, 'I want to do something for the molecule and the atom'. This spirit, I think, is one great asset America possesses over us.

There is one point, however, where I think the advantage is on our side and that is our system of secondary and university education. In every university I went to in America I was told by the scientific professors that hardly any of their students had any adequate knowledge of mathematics. It is difficult to find out what an American boy learns between, say, 14 and 18 years. At 14 he knows as a rule quite as much, if not more, than an English boy of the same age but at 18 he has dropped far behind.

They come away from school with a kind of vague interest in a good many things but they do not come away with trained minds and even those who are going to take science at the university have a very insufficient knowledge of mathematics. The university courses are governed by the theory that there is something undemocratic in distinguishing between the training of an honours and a pass man and an athlete. It has been tried but is not popular. It is contrary they say to democratic principles to make these distinctions. Democratic principles are mysterious!! I found that, though it is not democratic to distinguish between the training of a good mathematician and a poor one, yet no one thinks it undemocratic to distinguish at the university between the training of a good football player and a bad one.

Some explanation may be possible but it seems to an outsider that the one principle would cover both cases. They defend this system by saying it is good for the average man. I think there is a good deal of truth in this and that for the average man or rather for one rather below the average more is done by the American universities than by Cambridge. The research institutions, however, do not require the average man; for important posts exceptional men with good training are required. They do not get this type in sufficient numbers from their own universities, for they have no undergraduate training such as that given by the Second Parts of Cambridge Triposes specially adapted for such men. The result is the greatest difficulty in filling up places where a man has need of high scientific knowledge.

This is a matter where this country has a distinct advantage. I hope we shall stick to that advantage because, judging by what I see of the state of the industries in America, we shall need every advantage if we are to hold our own in competition.

It was generally wished at the Cavendish Laboratory to mark the occasion of J.J.'s 70th birthday, and it was decided to make the annual Cavendish dinner into an appropriate celebration. It was held on Saturday, December 18th, 1926, Dr Alexander Wood and Mr H. Thirkill being the organisers. Some of the party were entertained for the week-end at Trinity Lodge. These included Sir Richard Glazebrook, Sir Richard Gregory, Sir Oliver

Lodge, Sir Napier Shaw, Miss I. Woodward, and the present writer.

Cables and telegrams arrived from all parts of the world. The dinner was held at the University Arms Hotel, and there were 133 covers laid, nearly all for Cavendish Laboratory workers, past or present. Rutherford took the chair, with J.J. on his right. Lady Thomson came, and it was duly emphasised that she did so in her own right, as a past research worker in the laboratory. At the top table were Rutherford, J.J. Thomson, Langevin, Lady Thomson, Glazebrook, Schuster, Newall, Lady Rutherford, Threlfall, Lodge, Larmor. A pleasant feature was the presence of the laboratory assistants chiefly associated with J.J.'s régime—Everett, his personal assistant; Lincoln, the head of the workshop; Hayles, the lecture assistant; and Rolfe, who was the senior of them all, and had been associated with the laboratory as long as, or longer than, J.J. himself.

An address bearing the signatures of 230 of J.J.'s disciples, and with it two silver caskets, were presented by Rutherford to J.J. and Lady Thomson, and he spoke of his earliest days at the Cavendish Laboratory as 'the happiest in my life'. Prof. P. Langevin followed, as the representative of men of science from other lands, and J.J. replied. It was evident that he was much moved. The toast of 'The Old Cavendish Students' was proposed by Dr Alexander Wood, and Sir Arthur Schuster, Sir Richard Threlfall, and Prof. F. Horton replied. They were chosen as representing different epochs. It was a successful occasion. Sir Oliver Lodge's name was not on the toast list, but, as some of us half anticipated, he found an opportunity of speaking, and we heard him with pleasure then as always.

Next day (Sunday) there was a tea-time reception at the Lodge, which was attended by many of those who had been at the dinner. The address and presentation, together with other mementoes of J.J.'s career, were on view.

An amusing conversation in J.J.'s study during the Sunday is recalled. Sir Richard Gregory spoke of having had a book in for review—a compendium of mathematics by an Indian author.

Sir R. Gregory. 'On turning over the pages many passages

seemed strangely familiar; and I presently discovered the reason. Whole chapters were taken verbatim from books which I had read in my own student career. Algebra from Hall and Knight, Conic Sections from Charles Smith, and so on. The author seemed to think that a book was to be written by wholesale piracy from his predecessors.'

Another guest. 'Well, that is how it is done, only not quite so crudely.'

J.J. (with a chuckle). 'It seems to me that his only fault was one of technique!'

During the course of the 70th Birthday Celebrations, several messages of congratulation were received from overseas. The following from Prof. Alois F. Kovarik of Yale is chosen for reproduction here:

During the past as well as during the present generation, the Cavendish Laboratory has been the Mecca for physicists from all parts of the world, and we Americans owe a debt of gratitude to it for aiding and developing the spirit of research in physics in America. There is scarcely a physicist in America who has not been a pupil of Sir Joseph or else a pupil of one of his pupils. In the case of the latter, the enthusiasm for research and the high esteem for the Master came not only by reflection from those more fortunate ones who had been with the Master himself, but even in some cases with added force, for the powers inherent in his pupils were released by virtue of their contact with him. To these men Sir Joseph was made known not only as a physicist who opened the fields of electronics and of atomic structure, but also as a most congenial and sympathetic man. To them also he soon became simply 'J.J.' They seemed to learn to know him intimately even before meeting him personally on his several welcome visits to America.

Those of us who have been more fortunate and to be privileged to associate with him at one time or another have had an increasing esteem for him not only because of the great scientific achievements which his mind made possible, but also for those things which a man likes to find in another man. Not only did he show interest in one's scientific work, but also he always showed human interest in the man himself, his friends, and the institutions with which he was connected. We all love his characteristic smile, and every one of us felt a certain pleasure within ourselves on hearing a footstep that every Cavendish man recognises as solely 'J.J.'s'.

It was my good fortune to have been a pupil of two of his pupils and to have experienced the delights from such associations, but at a more recent time I had the privilege to be associated with him in the Cavendish and in Trinity—as a guest in both—and it is a pleasure to admit that my love for him, for his human and personal qualities, causes no less pleasure to me than my great admiration for his genius as a physicist.

Thomson was, as he himself records and as others observed, exceptionally helpless with his hands, and excepting a little of his earliest experimenting he relied on an assistant to do all the preparatory work. He describes how in his youth he nearly lost his eyesight in an explosion, but that is perhaps the exception which proves the rule. Later on at any rate he did hardly anything with his hands except to use the pen. He wrote an excellent handwriting and professed himself unable to dictate—a curious difficulty in so ready a speaker. In his later years his daughter, Miss Joan Thomson, helped with his correspondence, and kept his papers in order.

J.J.'s faithful assistant, Ebenezer Everett, was compelled to retire in 1930 on account of a breakdown in health. He suffered from heart weakness, combined with painful asthmatic attacks. 'I do indeed thank you', he wrote, 'for your great kindness to me. I have one disappointment, that is to leave you before you give up the laboratory. It has been my ambition to be with you as long as you were there.'

He was then sixty-five years of age, fifty of which had been spent in the service of the University, and forty-one in the Cavendish Laboratory. It was estimated that he had made upwards of 5000 pieces of special apparatus for the workers there. Only a small minority of the research students were much use at this kind of work, and Everett was nearly always called in to supply their deficiencies and to teach them. He was not always patient, however, when his work was destroyed by rough handling, and he was called upon to do it all over again. Owing to the exceptional value of his services he was given the recognition of an honorary M.A. degree, which has been very rarely given in such cases. It was the proudest day of his life when he walked from the Senate House accompanied by his beloved chief, by whom he was generously pensioned. After three years of invalid life he died in 1933. Thomson, Rutherford, Aston, and others for whom and with whom he had worked, attended his funeral. He was a vigorous Conservative in politics and a member of the West Chesterton Ward committee.

From the late R. S. Willows, on the occasion of his being elected a director of Tootal, Broadhurst, Lee, and Co. of Manchester, in recognition of his developing a method of making creaseless cotton fabric for the firm, which has become of great commercial importance:

Penwick Bay Hotel, Isle of Man.
15. 8. 1932.

Your kind letter of congratulation followed me here. Among the many I have received it was one of the most welcome, as you first awakened in me the research spirit.

Our success is the more pleasant from the fact that we were all *pure* scientists when we started on the problem, and were looked at askance by men in the trade, and both Mr K. Lee* and I were most anxious to show Lancs. that science is worth while.

We've had an excellent three years in the lab. and the knowledge gained in the earlier years has constantly borne fruit.

It's amusing to think I left the village school at 14 and was on my father's farm until nearly 19, and once resolved I would avoid Lancs. However, I am glad I entered industry as I find it very interesting, though when thousands have to be spent at my recommendation it's a worry.

In Cavendish Laboratory days, circa 1898, Willows was a quiet unobtrusive man who had not strongly impressed his qualities upon other research workers there. I asked J.J. if he had realised them, and he replied, 'Well, I knew he was a sticker'. This was one of the qualities J.J. valued most highly.

* Sir Kenneth Lee, now Chairman of the Company.

CHAPTER XIV

VIEWS ON EDUCATION

DURING the latter half of his life, J.J. was called upon on many occasions, such as prize giving, educational conferences, opening of new laboratories, after-dinner speeches and the like, to give his views on education. These pronouncements make possible a general reconstruction of his views, which will perhaps be better suited to this book than any summary of what he said on this or that occasion.

The general keynote is that he was entirely opposed to any kind of pedantry or formalism in education. Given an intelligent teacher and a small class he was inclined to favour an unconventional method more than a conventional one. In particular he wished to encourage educational experiments. He thought, as I believe do most other people who have considered the subject, that the modern tendency is to attempt to teach far too many subjects, and too much of them. He quoted a pathetic note written by a candidate at the foot of a science and art department examination paper. 'I am afraid I have done very badly, but this is the fifteenth subject I have been examined in in a fortnight, and my head is all in a muddle.'

It may be worth while parenthetically to consider what is the reason for this tendency which J.J. deplored and which continues in the face of general disapproval. I think it is rather like the reason which makes public expenditure constantly grow, in spite of the wish of taxpayers and ratepayers to keep it down. People lament general extravagance of government, whether central or local; but when it comes down from the general to the particular, they deplore, and do their best to defeat, its short-sighted economy about the special thing in which *they* are personally interested. Similarly in education; the specialist who is teaching one particular speciality, say human anatomy, or organic chemistry, or geometrical optics, gets very familiar with the elements of the subject, so familiar that it ceases to interest him, and he ceases to see the difficulties of it. He feels a keen interest in more recondite and probably less useful developments, and is pardonably anxious to interest his pupils in what interests himself. This leads to a constant extension of syllabuses and curricula; and no doubt the progress of knowledge makes some additions necessary. It is not enough borne in mind that the capacities of pupils do not stretch proportionately with the increase of knowledge, and that if new matter is to be included, some old matter must be jettisoned to make room for it. This is sure to be opposed by some of the old school; and there is a tendency to compromise by including the old and the new as well.

J.J.'s general attitude on this matter is amusingly illustrated in an article he wrote on Lord Kelvin.

In 1907, the last year of his life, Lord Kelvin said, 'A boy should have learnt by the age of twelve to write his own language with accuracy and some elegance. He should have a reading knowledge of French, should be able to translate Latin and easy Greek authors, and should have some acquaintance with German; having thus learnt the meaning of words, the boy should study logic.'

To retain such views as these at eighty-three must either mean an unconquerable optimism or else that the number of boys under twelve with whom Lord Kelvin was brought in contact was very limited.

J.J. considered that elementary science was taught to young boys far too much in the way that would be appropriate if they were eventually to become specialists in the subject, which of course the large majority of them were not. He thought that an easier and lighter treatment of the subject, particularly with reference to what may be seen in everyday life, would for most boys be of much more educational value.

J.J.'s pronouncements at various times as to what could be achieved by education were not always consistent. He sometimes took the line that very few people were stupid, and that if they seemed difficult to stimulate, the fault probably lay in the school system, which had failed to find the right subject to interest them. At other times he seemed less optimistic. In conversation he propounded as a summary of the theory of education, that you could not make a silk purse out of a sow's ear. He perhaps did not mean either thesis to be pressed too hard. He said in public that:

There was a kind of scheme advocated fitted for some abnormal structure called the average boy, and they seemed to regard the boy as a kind of clean piece of canvas on which any picture to be painted depended only on the skill of the artist. A truer representation to his mind would be that boys were very much more like a photographic plate on which a latent image had been formed, and the process of education consisted in a development of that, in bringing up the strong points and in the obliteration of the weaker ones.

Perhaps this really best represents his considered opinion on the subject.

He was very strong in maintaining the educational value of hobbies such as carpentry, photography, and the like, when boys had to overcome the difficulties which they encountered for themselves, and he was inclined to favour day schools over residential schools because they often gave better opportunities for this, and also incidentally because boys get the opportunity of learning something about social and economic questions at home, and generally of getting into an atmosphere where different conventions prevail. It also gave parents some scope for exerting influence. His advocacy of mechanical hobbies is the more striking because his own education, so far as can be judged, owed little to them. He was not apt with his hands, and never became so. He thought, however, that these occupations would go far to develop the capacity and self-reliance of boys who were not naturally bookish —and in some moods he seemed rather to put this sort of knowledge higher than academic knowledge. He remarked that if he had to teach a class of boys the elements of physics, he would begin with the motor bicycle. He thought that in this way the attention of boys would be secured in a way that would carry them through the initial intellectual difficulties. I once quoted to him an exclamation of Charles Rolls, well known in connection with the Rolls Royce car, who was an enthusiastic practical mechanic, and was up at Trinity as an undergraduate about 1896. 'What', said Rolls, 'is the use of talking about Cos θ if you cannot use a spanner?' J.J., who was certainly much more at home in manipulating

Cos θ than in manipulating a spanner, seemed nevertheless rather to applaud the sentiment, and said reasonably enough that he would rather be taken up in an aeroplane by a spanner expert.*

In the same spirit was a story J.J. used to tell of a candidate for the army entrance examination, whom he had examined in practical physics. Some electrical measurement had to be made—Wheatstone's bridge or the like, and the candidate had not the slightest idea how to do it. Undeterred by this, he tried joining the wires in every conceivable way. J.J. watched this proceeding with approval, and reported that though the candidate knew nothing of physics, he would be an uncommonly good man to be with in a tight place. What view the Army authorities took of this recommendation is not related.

As regards public elementary education, he was not particularly enthusiastic about raising the school-leaving age. He thought that the children were often far more interested in practical things such as the work taken up after leaving school, and that many of the most intelligent did not develop until this stage was reached.

J.J. occasionally made remarks about literary and linguistic studies, but it is difficult to extract any consistent body of doctrine from them. Speaking in 1912, he said that

It was always rare, and was now extremely so, to find a student who could translate a straightforward bit of German into English. In the old days, if a boy was set to learn German, the first thing he tried was translation, but now a man had to be most efficient in a foreign language before he could read a sentence of it. The modern method was one of teaching by sound, and he had suffered from it because it failed; when a boy he had practically no ear. When he studied German he had a teacher in advance of his time. This gentleman believed in teaching by sound.

He used to shout out a word. I shouted something I thought was like it. He said it was not. He repeated it. I could see no difference, but thought it incumbent on me to make a change. We went on shouting louder and louder to one another till at last we had to agree to differ. I do not ask that boys from school should be able to speak or even to write German, but they should be able to read it.

* However, poor Rolls' skill with the spanner did not save him from losing his life in this way.

As regards the classical languages, and classical education, it is difficult to know what he really thought. He had not had much classical education himself, and had got through the necessary minimum of Greek for the 'Littlego' by 'cramming', without, as he himself said, learning anything of the subject. If he had any want of sympathy with classical studies, he could not well have given expression to it when part of his audience were teachers whose whole *raison d'être* was bound up with these studies. Nor would it have been sympathetic to have expressed himself in this sense at Trinity. He did not fail to note that some leaders of industry preferred classically trained men as executive or administrative officers: and without denying that this might be justified, he thought the interpretation was ambiguous, so long as the ablest boys were pressed by the schoolmasters to remain on the classical side.

He thought that it was a defect of the public school system that the entrance scholarships had in practice the effect of attracting the able boys to classics. In the examination for most of their scholarships much greater weight is given to classics than to any other subject, and a boy must have spent most of his time on classics if he is to do well in the examination. Thus, when he goes to school he is much farther advanced in classics than in anything else, and naturally takes it as his main subject. It may not, however, be the subject in which his strength really lies: for unlike mathematics, in which marked proficiency is only attainable by boys with a somewhat rare type of mind, in classics most able boys can under skilful teaching become proficient enough to give them a fair chance of getting an entrance scholarship at a public school. These scholarships may then entice them along a path which does not lead to their true destination. That this actually occurred was shown, he said, by the evidence given before his committee of 1918. Of the entrance scholarships to Cambridge gained by boys from seven great public schools which give entrance scholarships, for one gained in science, six were gained in classics. This disproportion was far greater than for all schools, showing that it was not due to the rarity of scientific talent as compared with classical, but was an artificial one due to the system in force at these schools.

He said, however, that the last thing he wished to do was to disparage classical or literary studies. He thought that for some boys a course in which classics predominated was the best, and that in the early stages of education it should always play a large, perhaps even the largest, part. What he thought desirable was that the school examination should not be so much specialised, and that the papers in classics should not be so much more advanced than those in any other subject.

I do not remember during a long intimacy with J.J. that he ever used a classical illustration or allusion, or betrayed that he had heard of any personality, real or mythical, in antiquity, apart from the Greek mathematicians. My wife and I once took him and his family to see the remains of the Roman Wall in Northumberland, but though he seemed to enjoy the excursion regarded as a picnic, I do not think he asked questions about the antiquities, or betrayed interest in them. He was certainly anxious for the removal of compulsory Greek from the requirements for the 'Littlego', which made a minimum of Greek a necessary qualification for a Cambridge degree. I myself received a 'whip' from him when the question was to be voted on by the Senate in 1905.

The available facts are now before the reader and he can judge for himself. I think we must admit that if J.J. sometimes did lip service to the advantages of classics, he was not really a very effective or wholehearted advocate for them. I cannot recall that he ever advocated them when he was in the company of scientific men. What he really did hold about literary studies was that a boy should have such a command of English as would enable him to explain what he did and what he saw in carrying out an experiment in the plain English of educated people, and without making use of the conventional jargon of laboratories.

In the early days of the war of 1914–18, a 'Committee on the neglect of Science' was formed, with Sir E. Ray Lankester, the well-known zoologist, as chairman. A volume of essays was published on 'England's neglect of Science', and a League for the Promotion of Science in Education was formed. In June 1916 a deputation of this League, consisting of the late Lord Rayleigh,

Sir Ray Lankester, Dr Shipley,* Prof. A. G. Bourne, of Oxford, and Mr M. D. Hill, of Eton, were received by Lord Crewe. There is no doubt that the occasion was opportune, for the course of the war had forced to the front the imperative need for a better understanding of the scientific point of view, and a better supply of scientifically trained men, for all kinds of national purposes. The lessons of the war had to some extent woken up the official classes to a consciousness that science existed, and had national importance.

As a result of the deputation the then Prime Minister (Asquith) appointed a committee in August 1916:

To enquire into the position occupied by natural science in the Educational System of Great Britain, especially in Secondary Schools and Universities, and to advise what measures are needed to promote its study, regard being had to the requirements of a liberal education, to the advancement of pure science, and to the interests of the trades, industries and professions which particularly depend upon applied science.

J. J. Thomson was the chairman. The choice was an obvious one, both on account of his personal position, and his official position as President of the Royal Society. No information is now available about what may have passed between J. J. and the Prime Minister. J. J. used to say that he had found Asquith's general attitude towards science unsympathetic; it may or may not have been in connection with this particular matter.†

The report was drafted by a sub-committee consisting of J. J. Thomson, Sir Graham Balfour, Mr D. H. Nagel, Mr W. W. Vaughan, with Mr F. B. Stead the secretary.

The report of the committee is of course largely composed of statistical matter, and so far as J. J.'s own sentiments can be traced in it, they had been perhaps more amply expressed elsewhere, so that there is no occasion to dwell on the report at length. The keynote of the recommendations was of course that more time

* Afterwards Sir A. E. Shipley, G.B.E., Master of Christ's College, Cambridge.
† I feel bound to say that, as the result of several conversations with Lord Oxford at different times, I did not form the same impression.

and more money should be spent on science in all general education from the age of twelve years upwards.

The elements of natural science was to be a necessary subject in the entrance examination to the public schools, and due weight was to be given to this subject in the entrance scholarship examination to public schools. In the School Certificate Examination all candidates were to satisfy the examiners in both mathematics and natural science.

None of the main recommendations appear to have produced any effect in practice, and it is to be feared that any effect the report might have had in these directions must now be considered to have been exhausted. In fact, the whole incident is an excellent example of the usual official technique of shelving a question by the appointment of a committee to enquire. Naturally nothing could be done until the committee had enquired and reported. Seventeen eminent gentlemen were approached by a prominent politician, who invited them to serve their country as they were so well qualified to do by joining in this important investigation. They consented, and devoted intermittently some forty-five days to the work. Fifty-two witnesses—people whose time was of some value—were examined. An elaborate report was drafted, and after the exercise of considerable tact and address by the eminent man who had been appointed chairman, it was signed by all the members about eighteen months from the start. It would, of course, have been unreasonable to expect any immediate action, when time was required to digest the recommendations, and when a new Prime Minister had succeeded the one who appointed the committee—a Prime Minister who by the way had just then an exceptional share of the world's affairs on his shoulders, and had to be away in Paris for a great part of his time negotiating the Peace Treaty. After about two years more—four years in all—some of those who originally moved in the matter became impatient and proposed another deputation to the President of the Board of Education—and so it went on until the original enthusiasm of the movement was worn down by the almost Oriental system of saying Yes when you mean No.

J.J. does not seem to have been specially impatient of being treated in this way. It was hinted at the beginning of the report that something of the kind might happen.

Just now [it was stated], everyone is prepared to receive Science with open arms, to treat it as an honoured guest in our educational system, and to give it of our best. Just now, it seems almost unnecessary to take action to ensure against any relapse into the old conditions, but experience of the past shows us that temporary enthusiasm needs to be fortified by some more binding material.

These anticipations proved only too well founded. It was a case of

> When the Devil was sick, the devil a monk would be,
> When the Devil got well, the devil a monk was he.

During the time when the Royal Commission on the Universities was sitting and much criticism flying about, J.J. at a London dinner sat next a prominent Labour Member of Parliament who said: 'Professor Thomson, you need not be afraid of what Labour will do. No one has so high an opinion of the Universities as those who have never been there.'

J.J. never seemed to me specially interested in making the educational ladder easy to climb. He was not very pleased to see his laboratory attendants attempting to become University graduates, fearing apparently that they would find themselves in a false position if they succeeded. He often said it was a mistake to have a scholarship system that enabled a number of very poor students just to make ends meet at the University. If a man were to come up to Cambridge his funds should enable him to take full part in ordinary University activities, join one or two clubs, and entertain friends to a reasonable extent.

The successive crises in the University about the position of women will probably seem like a series of storms in a teacup when viewed in comparison with the great changes in the status of women which have been brought about by world events. However, they were important to J.J. and to the circle in which he lived and worked. He was sympathetic in the early days, being a close personal friend of Professor and Mrs Henry Sidgwick, who were protagonists of the movement for the University education of women in Cambridge. He opened his lectures to women students in 1885, and, presumably in recognition of his sympathetic attitude, was made an Honorary Member of Newnham College. The proposal to grant titular degrees to women was lost by an overwhelming majority in 1897. J.J. voted with the minority, though he does not seem to have entered the lists of public controversy.

In 1920 the question of women's status at Cambridge came up again, and the proposal was made to follow the example set by Oxford, and allow them to become full members of the University. This would have carried with it a vote on questions of policy, and a seat on Boards of Study. J.J. wrote to *The Times** to oppose this. He urged that much of the benefit of a university is lost if it consisted entirely of honour students, and that a university for women was needed which would cater for those women who did not look forward to any vocations other than those associated with home life: that Cambridge had no accommodation to provide for a large increase of this kind; and that if more were done for women honour students at Cambridge, it would make more difficult the foundation of an adequate university for all classes of women students elsewhere.

He deprecated giving women graduates a vote on the policy of the University, or a seat on Boards of Study, urging that this would raise difficult questions as to the differences in the regulations for men and women.

As might be expected, this letter did not commend itself to the heads of the women's colleges, Miss Jex-Blake and Miss B.A. Clough, and they wrote in reply, urging that it was against the spirit of the times, and pointing out that Oxford, London, and the newer universities all admitted women to pass degrees.

J.J. wrote in reply. But one can hardly help feeling that he was influenced not so much by the consideration of what was better or worse for women students or teachers, but rather by the fear that the University for which he had lived and worked might be changed out of recognition by the proposals before the Senate. His colleagues, Rutherford and W.J. Pope (Professor of Chemistry),

* Saturday, December 4th, 1920.

wrote in the opposite sense: but J.J.'s view prevailed, and although women teachers now sit on Faculty Boards, they are still not given degrees in the full sense, but only 'Titles of Degrees'. To most people this difference may well seem over-subtle, and it has not prevented the appointment of a woman as a University Professor.

J.J. Thomson was appointed in 1919 an original member of the University Grants Committee, by his old pupil Austen Chamberlain. This committee was to administer the financial assistance given by the Government to the universities, which the changing conditions of modern times, and particularly the requirements of scientific teaching, had made necessary. He was a pretty regular attendant during the all important early years when the policy of the committee was being shaped. By 1923 it had come to be decided that the University of Cambridge was to receive a regular grant of substantial amount, and he tendered his resignation on the ground that his position in the University made him an interested party. The then Chancellor of the Exchequer, Neville Chamberlain, agreed with this view, and his resignation was accepted.

On some occasions he expressed his views about lectures, and thought that the tendency of university students was to rely on them too much.

When a man studied from a book [he said], he could take his own time. At a lecture he had to adopt the lecturer's pace, which might be too fast for him, and in many cases the students took little trouble to understand what they heard. They put down as many as possible of the lecturer's words, trusting to discover their meaning afterwards. A textbook must be exceptionally bad if it was not more intelligible than the majority of notes made by students.... The proper function of lectures was not to give a student all the information he needed, but to rouse his enthusiasm so that he would gather knowledge himself, perhaps under difficulties.

He said that the absence of personal triumph in the acquisition of knowledge was really the thing to guard against in education.

If Thomson talked about examinations in any detail, it was generally to denounce the questions as too difficult. Thus, talking of the old mathematical tripos with order of merit, he said that examiners pondered the questions they were concocting too long. The question might be suitable enough in its original form, but after its author had turned it over in his mind for a few weeks, he had become so familiarised with it that he had ceased to see that it presented any difficulty. The next step was to put a twist into it to make it harder: and this progress might even be repeated, so that in its final shape the question became very formidable. This he considered led to time being taken away from the spirit of the subject in order to acquire a technical dexterity that was of comparatively little educational or practical value. It was this that led him to favour the abolition of the order of merit—abolishing the senior wrangler as it was popularly called. J.J. remarked on one occasion that that position often went to the swift rather than to the strong. If the candidates were merely classed instead of being put in order of merit, the edge would be taken off that keen competition which made it unsafe to ignore merely technical facility in unimportant matters. I believe that when it fell to him to examine, other mathematical teachers considered that his questions were too easy. One can readily understand that having spent time and effort in preparing their pupils to deal with hard questions, they were disconcerted when these were not set.

When examining for entrance to the Civil Service, he complained that the standard in physics was unduly high. He had to set hard questions as instructed, but that did not prevent him from considering the standard ill judged. This, it must be remembered, was long ago, and present conditions are very likely much altered.

He was also inclined to take the view that mathematical teaching was too formal. At Trinity, for example, during the early days of his Mastership, he pressed on the mathematical staff his view that they should rely less on formal lecturing and blackboard exposition, and do more by way of personal contact and exploring difficulties in conversation—in fact to approximate their methods to those used, for example, in the College teaching of history.

It is certain that this method would have very great value in the hands of a teacher like J.J. himself. I can remember to this day points of view which I gained from his comments on problems which he gave the class to work out 'as an example for next time'. He did not always remember that he had proposed these questions, and few members of the class offered any solution, but those who did were well rewarded by the illumination which they got from his comments. The mathematical staff at Trinity, consisting at that time of pure mathematicians, were not disposed to accept his advice, but it seems that the course of experience has justified it, and that the present system is more or less of the kind that he wanted.

He did not seem to take much interest in the questions about rigour that now occupy so much of the attention of mathematicians—all he cared about was that the answer should be right, and he held the view that very often so much time was spent in insisting on the limitations of useful mathematical methods that students never learned to use these methods with facility. More was made of the exceptions than of the rule. For example, he advised me while I was doing post-graduate work in the Cavendish Laboratory to gain some acquaintance with subjects like Fourier's Series, Spherical Harmonics, and the like. A little time afterwards he asked how I was getting on. I complained that to my taste too much of the book which he had recommended seemed to be given to examining the convergency or otherwise of the series used. J.J. said, 'You had better skip that.' 'Skip it altogether?' I asked. 'Yes,' he said. 'People spend all their time on convergency, and never learn how to use the series.'

No doubt mathematicians will be scandalised by this—but I repeat his words as I remember them. He cannot have meant that it was safe or possible to use series of unexplored properties without examining their convergency.

While J.J. was critical of many established educational methods, mainly on the ground that they had become too set and formal in character, there was one matter on which he spoke positively with no uncertain voice, and that was the educational value of research. The reader must here bear in mind a distinction. J.J.'s own life had largely been spent in research, and it was to his achievements in this direction that his fame and success were mainly due. His advocacy of research made it easy for any critic to compare him to the cobbler who says that there is nothing like leather. But to criticise him in this way would be to misunderstand his meaning. He did not assume that any great proportion of those whom he encouraged to attempt research would succeed in making important contributions to knowledge—and indeed his experience must have contained ample material to prove the contrary. As he himself said:

When I speak of students spending, as a means of Education, a year or so at research after taking their degrees, I do not contemplate that all or even any considerable proportion of them should adopt research as the business of their lives. To be successful at research and also to be satisfied with the rewards which a career of research has to offer, requires qualities which are not common, and which are of such a kind that unless a man is born with them he is not likely to acquire them in after life.

The special value which he attached to research as a means of education was that it necessarily took the learner away from that reliance on teachers to which all set teaching was subject. A man who was attempting even the most modest piece of research had to find out what others had done on the same lines, and he had to find it out for himself in the largely uncharted country of original literature, instead of having it presented to him in a cut and dried form by a lecturer. It is certain that many even among those who are successful in professional life do not know how to extract information from books, and indeed never dream of attempting to do so, and an educational system which has failed to give them any skill in doing it is open to criticism. But this was far from being all. J.J. maintained that he always saw the minds of those attempting research strengthen and mature under the process. It gave independence of view, self-reliance, initiative, and training in judgement: and the very disheartening phases through which a research worker generally goes before light begins to emerge was in itself a valuable training for the battle of life.

The teacher, he said, should not interfere more than was necessary to prevent the beginner from being disheartened by failure, and to prevent the work from getting on lines which could not lead to success. Not too much emphasis must be placed on the value of the results obtained, which could not in most cases be very great. In this respect he emphasised that the policy of a university laboratory should be altogether different from the policy of such institutions as the National Physical Laboratory, or the laboratory of a great firm.

To get scientific results rapidly [he said], the best plan is for the staff to select the subject for investigation, to determine the kind of experiment to be made, to exercise daily supervision over the work, and leave to the student little besides the taking of observations. The intellectual training of the student is injured rather than benefited by a training like this. You cannot without disaster apply methods of mass production to education.

It is to be feared that the application of J.J.'s methods and ideas is becoming increasingly difficult under the modern conditions of the growth of science. For example at the Cavendish Laboratory, where he fostered individualism so long and so carefully, the large plants for high potential work, the cyclotron, and the low temperature installation are obviously not to be run on individualistic lines. Team work is more and more replacing individualism.

Thomson was a good judge of men, and clearly distinguished their strong and their weak points, though his judgements were kindly. Of Rutherford he said many years after their first association, 'I saw his value at once'. When asked whether he had ever formed a similar judgement which had not been altogether confirmed by events, he admitted that he had in one case and gave the name.

Purely academic successes did not seem to impress him at all, or at least that was often his mood in his mature years. Thus, when he had expressed an unfavourable opinion of a Cambridge man who had received an important appointment, he was asked why, if so, the appointment had been made. He answered, 'Oh, just because he was a high wrangler. But he made mistakes which showed how little he was really good for.' He sometimes even went so far as to say that he was inclined to distrust men who were good at examinations. He made this remark in the presence of Lord Rayleigh, who was unable to digest it, and said: 'Well, Thomson, after all some of us have done pretty well in examinations.'

CHAPTER XV

MASTERSHIP OF TRINITY. CONTINUED

THE Mastership of Trinity could, without great paradox, be described as a sinecure, because the routine business of the College is mainly transacted by the Bursars and Tutors: but this is not the whole story. As chairman of the College Council, of most of the College Committees, and of the bodies which elect to scholarships and fellowships, the Master can exercise considerable influence: and therefore, though his statutory duties are not heavy, his responsibility is great. Since the influence he exerts is a personal one, depending more upon the man than upon the office, he can make as much or as little of his office as he likes.

So far as can be judged, it seems likely that history will regard Thomson's Mastership as a triumphant success. If any criticism could be made against him it was that he was not businesslike or methodical. He was a fairly punctual correspondent in answering a letter if he knew how to answer it. If he was embarrassed as to what he should say he was inclined to let it answer itself. Such a case arose when a well-known continental scientist proposed a provisional contract of marriage between their respective young children.

It was believed that in earlier days Lady Thomson found it advisable to search the wastepaper basket in his room before it was allowed to be emptied, otherwise dividend warrants and other important papers were apt to be unaccountably missing. During his Mastership papers were still apt to be mislaid and letters left unanswered. It was generally felt, however, that these defects were far more than outweighed by his qualities.

He never employed a whole-time paid secretary, and professed himself unable to dictate letters. In the earlier period of his Mastership, Mr W. H. Hayles, the lecture assistant at the Cavendish Laboratory, used to come in and help to arrange his papers and pamphlets. Later, his daughter, Miss Joan Thomson, acted as his secretary and kept systematic letter books.

Traditionally the Master, who had always been in Orders until Thomson's time, had been closely associated with the chapel services. This tradition was not broken. In earlier years before his Mastership, J.J. was regularly to be seen at chapel services on Sunday evenings, entering rather late with his hood awry. In December 1918, shortly after his appointment as Master, he was called upon to preach the commemoration sermon in the College Chapel. He did not feel it was very much his métier, and hoped not to be called upon in this way again; nevertheless, he characteristically agreed to do it. Contrary to his custom in public speaking the sermon was fully written out beforehand, which perhaps shows that he was not altogether at ease in giving it. Dealing with the terrible losses to the college in the war,

There is no scale [he said] by which we can measure losses like these, of great and varied gifts, of high hopes of promise of great services to their country, and to civilisation—the seed that would have yielded a great harvest has been destroyed.

He went on to pay a tribute to his predecessor, Dr Butler, and to look forward to the part the College and the University had to play in the rapidly changing social conditions of the time.

As Master he attended the morning and evening services on Sundays, though not as a rule on week-days. Considering the small attendance, the considerable expense of maintaining the chapel choir became somewhat of a problem. After successive relaxations any compulsion on undergraduates to attend had been practically abandoned in 1905, and this policy was definitely ratified in 1913. Thomson often said that he thought this was a mistake and he would have been glad after the war to re-establish some mild form of compulsion for at least one attendance every week. However, he does not seem to have made any move in this direction, probably because he found there was no prospect of adequate support.

During the General Strike of 1926 a good many undergraduates took the place of strikers for the maintenance of essential services, especially in some places such as Hull, where Labour feeling was very strong. J.J. often referred to this. The undergraduates, he

Family breakfast at the Lodge, Trinity College, 1937

said, fraternised with the strikers, and stood them drinks. He thought the public-school undergraduate and the working-class man generally got on well together, far better than the 'intellectual communist' and the labourer.

Somewhere in Whitehaven.
3. 8. 36.

SIR J. J. THOMSON,
   Master of Trinity College, Cambridge.

Sir,

I can't help but write a letter of thanks for what the students of Trinity College did for me and the others at Kinmont Camp Bootle.

Long before I heard of any camp I thought a great deal of students, but, since I had my experience of Kinmont Camp, and I feel sure all the unemployed that were there, will think the same, I think *much more* of students now.

First thing in the morning, last thing at night, wet or fine, the students had always a smile, better still can I put it like this:—'It didn't matter how mad they looked, they always had a smile upon their faces.'

Don't think one of your students *gave* me your address, 'cause how I got it was, in a roundabout way.

Thanking you once again,
   I am yours,
         Commonly known as 'Monty'.

P.S. When the boat race comes off it will be 'UP FOR CAMBRIDGE'.

It was the duty of the Master to take the chair at the half-yearly meetings of the College Finance Committee. There were three London members of the Committee, and what follows is mainly based on the recollections of one of them, Mr C. I. Bosanquet. All sorts of problems came up, such as the conversion of 5 per cent War Loan, the slump in Home Rails, the abandonment of the Gold Standard and the depreciation of sterling. It was also the time when economists were confounding each other and the world at large with their rival explanations of the forces governing the rate of interest and determining the price level. It was J.J.'s delight to examine all these matters in turn with the peculiar shrewdness of the North-countryman that is pointed by prejudice. If London members were inclined to feel a certain professional confidence in their own point of view, he would delight in asking just the question which, as they afterwards felt, had exposed the weak point in their position. His shrewd questioning always ended in an explosive chuckle.

He had probably too critical and destructive a standpoint to have made a good 'estates Bursar', and he was not always consistent in his criticism, but his keen mind and long experience of men and things was very helpful to the committee's discussions. For example, in 1932–33 some of the members were very anxious that the College should increase its holdings in really good agricultural land. J.J. was very critical of this whole policy, because he was convinced that Britain with a predominantly urban population would never accept a policy which involved dear food, and that therefore agriculture would always remain a depressed industry. His criticism was useful in that it stiffened the test which each proposal for an investment in agricultural land had to pass before it was approved. Whether his judgement was right is still in the womb of the future. At the moment (1942) the townsmen are hungry, and farming is in favour.

As regards the various Tithe Acts, 1925–1936, he joined in the protests made, but without taking any particular personal part. He considered tithe rent charge an unsatisfactory investment for the College, and rather welcomed its replacement by Government Stock, though he agreed that the terms of compensation were far from generous.

To preside at the elections to scholarships and fellowships was an important part of his duties as Master, though he had already had long experience of this work before succeeding to that office.

Trinity Lodge, 1924.

To G.P. THOMSON.

We brought out our entrance scholarship list this afternoon, so that I have been busy to-day, it is a big job now as about 75 took Physics, and I had to read their papers. The best man in science was a boy of 16.

An incident at a fellowship election may here be introduced, following the account of an eyewitness.

An elector was explaining the merits of a candidate and said by way of conclusion: 'Mr X has really said the last word about this subject.' The Master replied: 'But isn't it perhaps Mr X's weakness that he is so much better at saying the last word than the first one?'

It is said that when he was quite an old man, he conceived an unfavourable opinion of a fellowship dissertation on a mathematical subject which had been well spoken of by a referee. He took it away with him during the interval between the morning and afternoon meetings of the electors, and came back having detected a mistake which seriously vitiated the work.

With the possible exception of Bentley, no previous Master of Trinity could compete with Thomson in intellectual distinction and certainly no previous Master had shown more devotion to the interests of the College. He never spared himself in its service, even when, as necessarily sometimes happened, he had to deal with matters which were quite outside his previous interests or experience. His relations with the Trinity Mission in Camberwell afford an illustration. The Rev. Percy Herbert, now Bishop of Norwich, was warden and vicar at the time when Thomson became Master of the College. The workers at the Mission knew nothing of Thomson except his scientific reputation and were very much afraid that the personal touch of the Master on the Mission might prove a thing of the past. They were relieved to find that this was a complete mistake. Mr Herbert found that the new Master regarded the work of the Mission as an integral part of the life of the College, and took care to be acquainted with its developments. He always welcomed Mr Herbert to Trinity, and made him feel that the College was trusting him to do a branch of its work faithfully. Moreover, his shrewd practical advice was always ready. His help went far to make it possible to re-establish the Mission in the life of Trinity after the war, and it is in large measure due to him that the Mission still flourishes.

This willingness to throw himself into every College activity which could make a claim on his time came to be recognised, and in consequence he enjoyed the complete confidence of the Fellows. As a chairman, his quality was somewhat variable. He sometimes allowed his attention to wander, and let a discussion stray beyond the permissible limits: but on other occasions he would bring a discussion which someone's eloquence had entangled in foolish ingenuities sharply back to the limits of tact and good sense, perhaps with a flash of forcible wit. He had an acute sense of the greatness of Trinity tradition, and tolerated no approach to pettiness or arrogance in the relations of the College to the rest of the University.

As we have already seen, he had very definite views of his own in matters of finance and investment, and was by no means always disposed to yield them in deference to the professional opinions which impressed his colleagues. After endless resistance, he was often heavily out-voted, but the event not infrequently showed that he had been right. If his colleagues had known how well his own private investments had flourished, they might have been more inclined to defer to his views.

He could on occasion be obstinate, even when clearly wrong, as his pupils had come to realise in the old days at the Cavendish Laboratory, and this on occasion provoked the Fellows into sharp or rude retorts. He always took such attacks admirably, and junior Fellows have been seen quietly apologising to him at the end of a meeting without embarrassment on either side. He once remarked to a friend, 'If people are rude, the right thing is to take no notice'.

His attitude towards artistic problems was also sometimes distressing to those who had such matters at heart, but it was frank, like everything else that he said and did. 'In aesthetic matters,' he once told the Council, 'I have only one rule, to shout with the largest crowd.'

Some of those who served with him on the College Council have told me that they gratefully remember his success in keeping that body in a good temper, thereby facilitating the smooth passage of business. His success was largely due to the fact that he never lost his own temper.

One factor which increased Thomson's influence in College administration was his ready accessibility. This was no doubt more the result of the informality that was natural to him than to any set policy. The College officers, when they wished to consult him on matters of business, did not find it necessary to write letters or ring up on the telephone to find out when it would be convenient for the Master to see them. They went unannounced to his study, and always found him ready to give them as much of his time as they wanted. They valued his shrewd advice; not least perhaps because his outlook was not over-academic.

Nor were the Fellows the only disturbers of his peace. From time to time he was called upon by mothers of undergraduates, who came, sometimes in tears, to complain that Mr X, their son's Tutor, had been so very unpleasant and severe, and was it not perfectly reasonable for her to ask that her son might remain in residence, even though he had failed to pass the Law 'Special' which, as everybody knew, was one of the most difficult examinations in the world. It could not have been easy to comfort the mother and at the same time convince her that Mr X had a most tender heart and was merely obeying a College ordinance: but Thomson could rise to the occasion, and doubtless perfected his technique by practice. He once humorously remarked that one of his duties as Master was to wipe away from the cheeks of mothers the tears which hard-hearted Tutors had caused to flow; and the service he did in this way was not perhaps quite so trivial as it may seem, for the multiplication of angry and discontented parents may be a considerable menace to the good name of a college.

It had long been the tradition for the Master to dine in Hall and attend the Combination Room on Sunday evenings, and he improved on it by always inviting the party in the Combination Room to smoke with him in his study. Dr Butler, with his rooted aversion to tobacco, could not have done it: and apart from this, the position of the new Combination Room adjacent to the Lodge made the move much easier and more convenient. It was a joke against J.J. with the ladies of his family and their lady guests that he took his party from the Combination Room through the small drawing room without stopping to take any notice of them. These evenings were found very pleasant, and their usefulness went beyond that, for they did much to break down the isolation in which Masters of Trinity had hitherto dwelt, an isolation which was very unfavourable to their influence in the College. J.J. was a good and lively talker, and as his memories went back a long way he had a fund of stories which he told with good effect. He would tell of a meeting with Joseph Chamberlain, who complained that his son, Austen, then an undergraduate at Trinity, was 'Such a dreadful radical'—of a famous 'poll' coach of bygone days who was overheard saying to a pupil: 'No, no, it was Judas, not St John, who betrayed him', and other stories which have found a place elsewhere in this narrative, and in his own *Recollections*.

Those who stayed at the Lodge as his guests had similar or even better opportunities on week-days. I can recall the pleasure of several conversations of this kind when I had him to myself. On one occasion we discussed the old controversy about the discovery of the planet Neptune, and the criticisms of Sir George Airy, the Astronomer Royal, because, as was alleged, he would not attend to what Adams told him, and in consequence the glory of the discovery was lost to British Science.

Self. 'It seems to me that Airy's position was very strong. When he got Adams' letter he wrote asking questions about points on which he was not satisfied—and he got no answer. What more could he have done?'

J.J.T. 'Yes, I think Airy came out of it very well, but all the same I hope the people I have turned down were not so much in the right as Adams.'

Self. 'It would be very bad luck if they were.'

He was fond of talking about psychical research, and was by no means out of sympathy with those who pursued it. His name appeared for many years as a member of Council of the Society for Psychical Research, and afterwards as a Vice-President. He has given an account of his views and experiences in this matter in his *Recollections*. The general tone of them is non-committal rather than ultra-sceptical. On the subject of dowsing he goes so far as to say that there is no doubt of the reality of the process of dowsing. Most scientific men would, I believe, scarcely endorse this, but no doubt it is a scandal that any doubt should remain about a matter so important and so well accessible to experiment.*

Thomson was a much less awe-inspiring figure than most of his predecessors had been, and certainly it was his wish to be on easy terms with undergraduates. As an example of this, he was asked to a meeting of a very informal and spartan undergraduate luncheon club, and his acceptance gave great pleasure, not only to the members but to himself. He drew the obvious conclusion that he was asked not because it was customary, but because his society was desired. In former régimes such an incident would have been almost inconceivable. His sympathy with young men extended also to children, and he was very fond of romping with his grandchildren, as he had done in earlier years with his children.

He was most solicitous in his enquiries about the health of his friends and those who were connected with them, and he followed with anxiety the bulletins which appeared in the press even in cases where he hardly knew the people concerned except by reputation. He was most ready with his congratulations when his pupils or friends had achieved any small success.

In the summer his favourite resort was Fenner's, and on one of these visits he sat next a former member of the College who declared that he had never been more astonished than when he discovered that the old gentleman who had shown himself so well informed about cricket was the Master of Trinity: and undergraduates who were being entertained at the Lodge were often surprised to find how much the Master knew about the athletics of the College. Indeed, he sometimes overshot the mark, assuming, quite wrongly, that the young men were as much interested in this subject as he himself was, when in fact their interests lay elsewhere.

He wanted Trinity men to excel in all athletic competitions and was by no means satisfied with their very moderate success in

* The experiments required to settle it should admit of satisfactory repetition, and should give a definite answer, yes or no, as to whether the dowser had succeeded in each test. E.g. a series of pipes might be laid underground, no secret being made of where they were located. Then the stream of water would be *secretly* turned into any one of them at pleasure, and the dowser could be invited to discover which pipe it had been turned into.

these encounters. If they did not do better, it was certainly not from lack of his encouragement. He was often to be seen on the tow-path at College races of quite secondary importance or even at practices, as keen and interested as the youngest undergraduate there, and taking the opportunity of conversation with the oarsmen afterwards. Not infrequently he was the sole Trinity representative at a football match with another college.

His fame as an authority on these subjects seems to have spread to a wider circle. The letter that follows is an interesting illustration of this and of how varied is the correspondence of a Master of Trinity. It was carefully answered, but history does not relate what the answer was.

10 *St James' Square, Bournemouth.*
March 21, 1930.

Dear Sir,

I am the only sister of Sir Robert Penrose Fitzgerald, for 25 years M.P. for Cambridge. Every year for the sake of 'Auld Lang Syne' I have betted on Cambridge in the boat race; as though I am fairly well, thank God, at present, still at my great age (83) I can't expect to live much longer; and I felt inclined to put rather a heavy bet on this time. Last year from watching the crews on the screen I felt the race was a foregone conclusion; (for I spent all my young days rowing my own boat on Cork Harbour) (so I am rather a good judge) but this year nothing has appeared on the screen so far; and the accounts in the different newspapers are most conflicting; I thought I would make so very bold as to write and ask you if you thought I would be safe in putting a substantial bet on Cambridge? It would not be fair to my heirs if I lost much over it. I heard of you from my dear, *dear* nephew, Maurice Penrose Fitzgerald who was at Trinity preparing to be a clergyman, and was killed in the war. (The deepest grief in my life.) He had a deep true affection and admiration for you. Of course any opinion you express on the subject I will keep perfectly private if you wish it. I must apologise very much for thus troubling you.

Believe me to remain very truly yours,
GERALDINE PENROSE FITZGERALD.

During the earlier part of his life, both as an undergraduate himself and later as a teacher of mathematics and physics, Thomson had mainly been in contact with undergraduates of the intellectual type. I am not sure, however, that these were really the most congenial to him. In earlier days he seemed even more interested in the personal idiosyncrasies and amusements of his pupils than in their work. For example, one of the men who was doing research work at the Cavendish Laboratory was probably unique among them in being fond of hunting and shooting; and J.J. having discovered this was always ready to talk about field sports with him. He seemed to know all about undergraduate friends of my own whom he can scarcely have met. In fact, so minute was his information and so little obvious was its source, that it almost seemed to suggest a sort of secret service. In reality he must, I suppose, have kept a close watch on athletic records, and have remembered every chance word about the young men that came his way. When he succeeded to the Mastership, this side of his activities became in some sense a duty rather than a hobby, while he found the society of young men not of the studious kind was a relatively new and unexplored field. He made no secret of his opinion that these might be and very often were of great value to the College, and likely to play a more useful part in after life than the well-trained boy whose ambition was to obtain a first class in his tripos and possibly a fellowship. He may have exaggerated the value to the College and the community of these light-hearted youths: but many of them must have greatly benefited from discovering that a great scientist was very human, and not so different from themselves as they had believed. It is typical of Thomson that when an undergraduate whom he knew, and who, being in danger of failing in the Natural Sciences Tripos, just managed to 'scrape a third' he was delighted, for he had feared the worst. 'He could not have congratulated me more warmly,' said the young man, 'if I had been awarded the Nobel Prize'; and the exaggerated congratulation was very characteristic. He liked the undergraduate in question, and therefore was genuinely pleased that he had escaped the mortification of failure.

It was this combination of great intellectual eminence with all-embracing human interests that enabled Thomson to be so successful as Master. Owing to his reputation in the scientific world, he was well qualified to represent the College outside Cambridge; but within the College that reputation would not have been enough by itself. To be successful as the ruler of a society which is far from being uniform in its composition, he had to be able to sympathise with many different and conflicting points of view, to be skilled in the handling of men, and to find his greatest happiness in serving the College; and in this he did not fail. His Mastership was a great and continually growing success.

The ample size of the Lodge, and the comparative affluence of a Master of Trinity, has made it traditionally a centre of hospitality, where distinguished men were entertained when they came for University or College functions. In particular, it is an established custom for the successive Chancellors of the University, who have for a long time past all been Trinity men, to stay there on their official visits. Lord Balfour and Lord Baldwin, the Chancellors during Thomson's Mastership, always did so.

As we have seen his appointment was made in war-time, and the University was almost depleted of undergraduates. There were in residence at Trinity a large number of army cadets in training as officers, and these were entertained at the Lodge after chapel on Sundays in detachments of about thirty. In May 1919, after the war was over, two dances were given at the Lodge for some naval officers who were sent to the College for a course. There is reason to believe that they appreciated the privilege of their stay in Cambridge very highly.

Some names from the Visitors' Book may be given, under the dates when they first came. The list is not exhaustive even of distinguished names.

1918.   Sir Austen Chamberlain; Sir Oliver and Lady Lodge; Mrs Sidgwick; Prof. Volterra; Sir George and Lady Beilby; Col. H. G. and Mrs Lyons.

1919.   Prof. C. E. Mendenhall; Sir Charles Parsons; Mr A. J. Balfour and Miss Balfour; Bishop Bernard (later Archbishop of Dublin and Provost of Trinity College, Dublin); General Seely (Lord Mottistone); Mr Bonar Law.

1920.   Lord Crewe; Sir Arthur Evans; Sir Frederick Pollock; Sir Horace Lamb; Mr Gerald and Lady Betty Balfour.

1921. Sir Frederick and Lady Sykes; The Crown Prince of Japan (now the Emperor); Prince Kan-in and their Suite; Prof. and Mrs S. Arrhenius.

1922. Lord and Lady Charnwood; Prof. and Mrs Zeeman.

1923. Sir Joseph Petavel; Sir William Bragg; Bishop of London (Dr Winnington Ingram); Prof. and Mrs H.A. Lorentz; Prof. and Mrs A. Haller; Prof. Theodore Lyman; Prof. Ivan Pavlov; Lord Ullswater; Dr Alington.

1924. Lord Hugh Cecil; Duc and Duchesse de Broglie; Lord Willingdon.

1925. Rev. Sir George Adam Smith; Mr John Rawlinson, M.P.; M.B. Baillaud; Comte A. de la Baume Pluvinel; Mr Stanley Baldwin; Sir Malcolm MacNaughton; Lord Darling; Lord Dunedin; Dean of Westminster (Dr Foxley Norris).

1926. Sir Joseph Petavel; Viscount Grey of Fallodon; Viscount Haldane; Lord Merrivale; Sir R. Glazebrook; Sir Napier Shaw; Lord Wright.

1927. M. Paul Painlevé; Lord and Lady Lytton; Lord Ronaldshay (now Marquis of Zetland); M. André Maurois; Sir Cecil Hurst.

1929. Dr A. Gordon; Mr Desmond MacCarthy; Lord Hardinge of Penshurst; Crown Prince and Princess of Sweden; Sir Martin Conway; Lord Warrington of Clyffe.

1930. Sir J.M. Barrie; Prof. and Mrs Max Planck.

1931. Mr J.H. Thomas; Lord Noel-Buxton; Earl of Crawford and Balcarres; Lord Lee of Fareham; Lord Blanesburgh; Bishop of Blackburn; Sir James Jeans; Sir Herbert Richmond; Canon and Miss Lyttelton.

1932. Sir Walter Runciman; Sir William Llewellyn; Lord Sankey and Miss Sankey; Lord Hartington.

1933. Lord and Lady Hanworth; Dr Zenneck; Marchese and Marchesa Marconi; Mr Owen Hugh Smith; Mr Charles Bosanquet; Mr John Buchan.

1934. Mr Neville Chamberlain; Bishop of Croydon (Dr Woods); Prof. and Mrs Karl Przibram; Lord Leicester; Lord Coke.

1935. Prof. and Mrs Arthur Compton; Archbishop of Canterbury (Dr C.G. Lang); Lord Finlay.

1937. The Emperor Hailé Sellasie of Abyssinia and his Foreign Minister; Bishop of Norwich and Mrs Pollock; Dean of Westminster and Mrs de Labillière; General Wavell.

1938. Lord Hunsdon.

Apart from the Annual Commemoration of Benefactors, there

were several rather elaborate College celebrations for which the Thomsons entertained. These were:

1924. Sexcentenary of the Foundation of Michael House (one of the parent constituent colleges of Trinity).
1926. Francis Bacon Celebration (3rd centenary of his death).
1934. Coke Celebration (3rd centenary of his death).
1935. Fisher Celebration (with St John's, Queens' and Christ's Colleges. 4th centenary of his death).

J.J. enjoyed the experience of making contacts outside the circles in which he ordinarily moved, whether these were the humble people whom he helped to entertain in earlier days (see p. 158) or the great ones of the earth whom he was called upon to entertain at the Lodge. He certainly did not over-emphasise the importance of these latter contacts. As an amusing illustration of this, when travelling with his daughter, he bought a picture paper for her, and she handed it back to him with the remark, 'Isn't that an excellent photograph of the Swedish Crown Prince?' 'I can't say,' J.J. replied, 'I don't think I ever saw him.' It was with difficulty that she persuaded him that the Prince had stayed at the Lodge a few years previously. He did not often criticise people without restraint, but I have heard him express the strongest dislike and contempt for a distinguished man whom he considered a toady of the great.

When it came to entertaining foreigners, he would not make the slightest attempt at speaking French or German, though of course he could read scientific papers in those languages almost as easily as if they had been in English. I remember being a fellow guest at the Lodge with some distinguished French astronomers who had no English. I struggled to find something which a very limited command of French would allow me to say. The Master listened without attempting to join in, until he heard something that interested him, and then he broke in in English, bringing my laboriously constructed edifice to the ground!

In earlier days when he was showing foreigners round the laboratory and they did not understand what he said in English, his rather naïve method was to say it louder and louder, until they capitulated. The result was positively deafening. He was equally resolute in not saying one word in French, even when visiting Paris to receive an honorary degree in 1933. He had to sit through some dinners completely dumb.

In addition to the distinguished visitors from outside, there was of course a good deal of entertaining of members of the College, graduate and undergraduate. Lunch on Sundays was often devoted to the latter. Tea parties were given from time to time in the Lodge for the bedmakers and helps (bedmakers' assistants). The tea was followed by some form of entertainment, musical or otherwise. Lady Thomson of course shouldered the chief burden of entertaining, but J.J. was a sympathetic onlooker. One of the Fellows was much amused when his bedmaker remarked to him, after one of these occasions, that she thought 'Lady Thomson was a real lady, for she treated the bedmakers of Whewell's Court and New Court* exactly the same as those of Nevile's Court and the Great Court'.

Thomson did not at any time hold the Vice-Chancellorship. This office is ordinarily held by the heads of colleges in rotation, but he judged that Lady Thomson's health was not equal to the duties that would have fallen on her, and excused himself on this ground.

* These are the newer and less dignified parts of the College.

CHAPTER XVI

THE CLOSING YEARS

J. J. Thomson's will was ultimately proved for no less a sum than £82,000, and seeing that he had begun life practically without capital and had never engaged directly in business, this seems to show remarkable skill in managing his investments. During the earlier part of his life he had only his salary as Cavendish Professor, and later, his Royal Institution Professorship. The stipend as Master of Trinity hardly covered the expense of living and entertaining in the style expected, and the only extra money he ever received was the Nobel Prize of some £7000 or £8000. Certainly his financial success was not achieved by any kind of pinching. He was generous and hospitable in his earlier as in his later days—entertaining his friends and pupils as lavishly as his circumstances reasonably allowed, and being very liberal with wedding presents and the like. Later, he would give £100 here and £100 there, when those who had claims upon him were in difficulties. He did his utmost for the unfortunate widows and dependents of scientific men who came to England as the result of the anti-Semitism in Germany.

As to exactly how his money was made, no very definite information is to hand. One on whose advice he relied was his brother-in-law, Mr J. H. Batty, who had been married to Lady Thomson's twin sister. With him he was on intimate and affectionate terms. Mr Batty was and is chairman of the Ashanti Goldfields Company, and other important enterprises, and Thomson would talk round the subject of these, gaining what information he could, and sometimes saying explicitly that he had some thousands to invest and asking for advice. He never told Mr Batty (nor apparently anyone else) what he had done in the way of buying or selling, probably feeling that to do so would be to fix an unfair share of responsibility on him. Mr Batty was as much surprised as anyone when it ultimately appeared how successful the result had been. He had no doubt, however, that Thomson must have had a good instinct for when to sell, not trying to hit off the exact top of the market, but taking a profit when he had made it, and not grudging the other man a reasonably advantageous bargain. It is certain that Thomson had a great natural aptitude for, and interest in, business. He was broad-minded and got hold of the essential points. He always liked to meet with business people when travelling. Thus on a cruise he spent hours talking to an apparently rather commonplace looking man who he said was most interesting, being engaged in the Yorkshire textile trade.

Nothing interested Thomson more than an anecdote about some magnate in the world of English or American finance. He was particularly struck with the way in which these men when engaging subordinates believe they can see at a glance, or at any rate after a little conversation, whether a man will be useful to them, while they are almost impervious to the evidence of testimonials.

Thomson was also fond of hearing or repeating accounts of how fortunes had been made in unexpected ways. There seems little doubt that if he had turned his attention in that direction he might have been a great figure in the financial world himself. Mr Batty's general impression was that he was just as much a business man as if he had given his life to it.

Some who were not at close quarters with J.J. imagined that he must have made money by patents, like Lord Kelvin, but this was not the case. The only patent he ever took out was for an automatic Töpler vacuum pump using sulphuric acid as the working fluid. It was developed in the Cavendish Laboratory but was never used there to any important extent, nor, so far as I know, anywhere else.

It is interesting to notice the resemblance of J.J.'s financial history with that of Sir Isaac Newton. Both men went up to Trinity in their youth with meagre resources, both devoted their best years to academic pursuits, both got valuable appointments in their maturity, and both had the financial ability which enabled them to leave considerable fortunes, of comparable amount.*

* I owe this remark to Prof. C. D. Broad.

From PROF. A. E. HOUSMAN (on the occasion of the writer's seventieth birthday):

Trinity College.   27th March, 1929.

My dear Master,

Many thanks for your kind and complimentary letter. I am very susceptible to comments on my personal appearance. One day when I was just turned 40, I was walking along and brooding on the fact, when a passing carter of some 25 summers said 'What's the time, young fellow?'

A spring of joy gushed from my heart and I blessed him unaware.

Yours sincerely,

A. E. HOUSMAN.

To MRS H. F. REID:

Trinity Lodge.   Nov. 8th, 1935.

I have been pestered into writing 'Reminiscences' and am sorry I gave way, it involves far more work than I ever imagined. I have got to hate dates, they take so much looking up.

From SIR AUSTEN CHAMBERLAIN (Chancellor Elect of Reading University):

Oct. 1st, 1935.

You will appreciate the delicious humour of your first pupil conferring a degree upon you, but you will also, I hope, realise my desire to offer you the only mark of respect and friendship which it is in my power to bestow, and the pleasure which it would be to associate the Master of my college with my acceptance of this academic position.

To G. P. THOMSON:

Feb. 17th, 1930.

I did not find Broadcasting quite so dull as I expected though I would rather face the largest audience than read aloud in an empty room, however I had Bragg and his daughter and Janet for an audience which helped a good deal.

This letter refers to a Broadcast lecture which is printed in *The Listener* for January 29th, 1930, on 'Tendencies of Recent Investigations in the Field of Physics'. Some extracts from this will be of general interest.

A great discovery is not a terminus, but an avenue leading to regions hitherto unknown. We climb to the top of the peak, and find that it reveals to us another higher than any we have yet seen and so it goes on. The additions to our knowledge of Physics made in a generation do not get smaller or less fundamental or less revolutionary as one generation succeeds another. The sum of our knowledge is not like what mathematicians call a convergent series, where each new addition is less important than the one which went before, and where the study of a few terms may give the general properties of the whole. Physics corresponds rather to the other type of series called divergent, where the terms which are added one after another do not get smaller and smaller, and where the conclusions we draw from the few terms we know cannot be trusted to be those we should draw if further knowledge were at our disposal. . . .

(He then illustrates this by the problem of the age of the earth, showing how the old ideas are upset by the discovery of radioactivity, and continues):

I think this is a warning against taking too seriously speculations about either the remote past or the remote future of the Universe; founded as they must be on the physics of the moment. . . .

There is now a school of mathematical Physicists which objects to the introduction of ideas which do not relate to things which can actually be observed and measured. Thus, before very high vacua were obtained, it would not have been legitimate to speak of the mass or position of a molecule, but only of that of a finite volume of gas. The atomic theory of chemistry and the Kinetic Theory of Gases would have had to wait until the technique of high vacua had been developed. A similar view was introduced into metaphysics long ago by Bishop Berkeley, who held that it was impossible to maintain that a quality existed unless one knew how to measure its magnitude. I believe that this view now gets no support from metaphysicians. I think it is bad Physics as well as bad Metaphysics. I hold that if the introduction of a quantity promotes clearness of thought, then even if at the moment we have no means of determining it with precision, its introduction is not only legitimate but desirable. The immeasurable of to-day may be the measurable of to-morrow. A striking example of this is that a movement was started by some chemists at the end of last century to give up thinking in terms of the atomic theory on the ground that the mass of an atom could not be measured. By the irony of fate the movement had hardly begun when a method of measuring the mass was discovered. It is dangerous to base a philosophy on the assumption that what I know not can never be knowledge.* . . .

* This sentiment was attributed to Dr Jowett in some satirical verses current in the Oxford of his day.   R.

The qualities we look for in a theory depend on the view we take of the state of development of the science. If we think that it is unlikely that any discoveries will arise which will, like those made in the last forty years, revolutionise our outlook on the structure of matter and physical phenomena, the most important quality will be its power of developing and co-ordinating the knowledge we already possess. There are some, however, and I am one of them, who think that what has been found is but a small fraction of what there is to find, that the electron and the proton are not the last words in the story of the structure of matter, that there are rays other than α, β, γ and Röntgen rays waiting to be discovered, that still, as Newton said, 'the vast ocean of truth lies undiscovered before us'. To these it seems that the increase of our knowledge by experiment is the most vital of all considerations, and that the most important quality for a theory is its power of suggesting new fields for research. They regard a theory as a tool and not a creed, as an instrument for directing research and not as something which it is a heresy to doubt, and are willing to accept suggestions from any theory which does not contradict known facts . . . .

Important discoveries have repeatedly been made by people finding something they were not looking for, and what seems a troublesome interruption may turn out to be far more important than the theory we were testing. There are to my mind indications that concentration on the theory is a tendency of modern Physics. As I think this point is important, I venture to illustrate it by a domestic incident. Many years ago, when I was living in college rooms, I happened to tell my bedmaker that at Oxford they had men scouts instead of bedmakers. She said she was sure if that were so, the staircases would be very dirty. In the true scientific spirit she determined to test her theory, and went to Oxford when next there was an excursion. The next morning when I saw her she was in great glee. She said, 'I was right, Sir, about the staircases. I went up every staircase I could find, and they were all much dirtier than ours in Trinity.' I asked her what she thought about Oxford which she had never seen before, and I found that the buildings, the courts and walks and the river had left absolutely no impression upon her. She had just regarded them as obstacles in the way of getting at staircases on which she had concentrated. She had proved her theory but she had missed Oxford.

To G. P. THOMSON:

Trinity Lodge.   Sept. 29, 1935.

I have lately been writing for the Reminiscences a very short account of Electronic waves and have again come across a point which has before caused me some uneasiness, and that is about the statement that the electronic waves guide the motion of the electron. This would imply that when a moving electron is acted on by an electric force, it is because the waves are deflected by the force and then drag the electron after them. Now some years ago I tried the experiment of sending γ radiation whose wave-length was of the order of that of the electronic waves for fast electrons between two parallel plates close together, and applied an electric field which would have been strong enough to drive the electrons against the plate and stop them coming through. At first I thought I got an effect with the γ rays. I spent a good deal of time in removing possible sources of error—and with the final form of the apparatus I got no effect.

On thinking it over it seems to me that the stream of electrons which travel along the diffracted beams of the electronic waves might be explained without supposing that the incident electrons were guided into these paths by electronic waves. These electronic waves are at very high frequency and the photons corresponding to them have large amounts of energy, comparable with that possessed by an electron at rest. Now the dynamics of the collision of electrons as worked out by Compton shows that in this case the photon loses a considerable fraction of its energy, which is transferred to the electron which moves off in the direction of the incident photon. Thus the diffracted photons if they meet with any electrons in their course through the diffraction grating would start them off at high speeds in the direction of the diffracted light. These electrons would not be the original electrons deflected by being guided by the electronic waves, but fresh electrons started by these waves. I should be glad to know what you think of this idea.

In his later years J.J. was of course often called on as an after-dinner speaker, both at Trinity and elsewhere. In this capacity he was somewhat unequal. At his best, he was as good as anyone. At his worst, he fell much below this standard, possibly because he had not given enough thought beforehand to what he was going to say, but trusted to the inspiration of the moment. He generally came up well to the mark on occasions of special importance, which makes this explanation the more likely. Thus Sir William Dampier writes:

When talking once to a Student of Christ Church it occurred to me to suggest that our respective colleges might form one of those alliances which have been successfully established between other pairs of Oxford and Cambridge houses. After some delay due to a lapse of memory, the Master and the Dean of Christ Church (Julian White) arranged the alliance and an invitation arrived for the Master and six fellows to attend a Christ Church 'Gaudy'. I was one of the six chosen, and I think most of the rest were very anxious about the kind of speech the Master would make. We need have had no fear; he made one of the best after-dinner speeches I have ever heard, and delighted both his Cambridge colleagues and our Oxford hosts.

On this occasion he asked his colleagues to forgive his travelling in a different railway compartment, in order that he might have quiet to think over what he would say. A few days later, at the Annual Trinity Commemoration, his speech was too long, and far from successful.

One or two examples of J.J.'s style of after-dinner speaking may here be given. Thus, on May 18th, 1912, he attended a dinner of the Fisher Society, a Roman Catholic Society chiefly of undergraduates with undergraduate officers. Cardinal Bourne was the chief guest, and J.J. responded to the toast of Literature and Science. In the course of his reply he said:

The greater part of the science with which he was concerned dealt with the structure and properties of matter. He was sorry to say that frequently he heard matter spoken of with some, might he say, disrespect—he was fortunate if he escaped a sermon without a reference to 'mere matter' as if matter were almost beneath the notice of the preacher. He ventured to think that attitude was foolish. Matter was one of the most mysterious and entrancing things that could exist. If we could only see what was in matter we would see that the smallest speck of the meanest matter was full of systems whose numbers were as the stars in infinity; we should see the most wonderful processes going on. If we knew the mystery of matter the whole social conditions of the world would be changed. The part that science had played in lightening the toil and alleviating the suffering of humanity at present would be nothing to what would be done if we could only fathom the secret of the mystery of matter, and it was to the elucidation of that secret that men of science had devoted their lives. Their goal was far, far away; they had only started to approach it. Whether the race would ever approach it or not was, he thought, a matter quite open to question, but at any rate in the last twenty years they had seen an advance in that direction which had been quite remarkable, and a thing which he thought the historians of this period would have to acknowledge as one of the features of this epoch.

J.J. sometimes introduced unusual quotations into his speeches. Thus, when he was speaking of the cleaning and redecoration of the front of the Great Gate, he used

> Evidence of a disordered mind,
> Clothed in front, he leaves him bare behind.*

When a speaker had failed, and a substitute was happily found at short notice, he sought for a suitable quotation from the Bible about people bringing good news, and at Miss Joan Thomson's suggestion, he used

> How beautiful upon the mountains are the feet of him that bringeth good tidings.

An incident which caused a good deal of amusement was as follows. A paragraph had appeared in the *Manchester Guardian*, representing some one of prominence in local government there as depreciating the value of book learning.

> There was (he said) a clever boy at school with me, little Joey Thomson, who took all the prizes. But what good has all his book learning done him? Who ever hears of little Joey Thomson now?

Mr Owen Hugh Smith, then Prime Warden of the Fishmongers' Company, told this story at a banquet where J.J. was the guest of honour. J.J., rising to reply, said:

> I wish you were *not* going to hear little Joey Thomson now!

J.J. Thomson unveiled in Westminster Abbey the tablet to the memory of Rayleigh in 1921,† and the tablets to the memory of Faraday and Clerk-Maxwell on September 30th, 1931, the occasion of the centenary of Maxwell's birth. The following passage from his address is worthy of preservation:

---

* I have not found the source of this quotation. Some readers will doubtless be more successful.
† For his speech on this occasion see *Life of Lord Rayleigh* (London, 1924) by the present writer.

The work of these two men affords the most striking example of the influence which researches in pure science can have on our industries, on the amenities of life. There has been ample demonstration of this during the Faraday Celebration for Faraday's discovery. Maxwell's discovery led to wireless telegraphy, this too has given rise to a great industry, and has potentialities, political, social and educational, which we are only beginning to realise. The point I want to emphasise is that these consequences have resulted from researches which were made without any idea of any practical applications or financial results.

If they had started with this object they would never have made these discoveries, and new industries would not have been created. If Faraday had studied with the intention of improving the means of obtaining electric currents, he would doubtless have improved voltaic batteries, but he would not have discovered the dynamo. If Maxwell had worked with the intention of improving communication between people at a distance, he might have improved the ordinary telegraph but done nothing that would have led to wireless. These great discoveries, which create new industries and revolutionise old ones, are not made by pursuing the obvious. In these days when the problem of unemployment lies so heavily upon us it is important to keep in mind what the patient search for scientific truth without any thought of industrial application has done in creating new industries and thereby creating employment.

There is a well-known bust of J. J. Thomson in the library of Trinity College by the late Derwent Wood, R.A., and the circumstances under which it came to be made are of some interest. I owe the following particulars to Mrs Derwent Wood and to Mr A. W. Ganz. Derwent Wood met Sir Joseph Thomson in 1919 or 1920 in connection with the memorial to the late Lord Rayleigh in Westminster Abbey, upon which he was engaged, and was greatly struck by the sculptural qualities of Thomson's features. He was anxious to get an opportunity of doing his bust and the opportunity came unexpectedly soon. The Cambridge Professor of Fine Arts approached Derwent Wood one day and asked him to come up to Cambridge and give a demonstration of modelling to the students. Derwent Wood consented, provided the Master of Trinity would sit as model. J.J. at once agreed and invited the Derwent Woods to the Lodge. After dinner they went round to the Architectural School where the students had gathered in keen anticipation, and the Master took his seat on the throne. Derwent Wood worked very quickly, and the bust was finished in an hour. There was a little speech-making at the end, and J.J. said:

> I was intensely interested in watching you at work and extremely relieved. I didn't know busts were made in this fashion. I was expecting any moment you would take an impression off my face with the clay.

(That was no doubt how J.J. would have done it himself if circumstances had compelled him to make the attempt.)

Later J.J. came to Derwent Wood's studio for two sittings, but he thought no more about the matter. In the meantime Mr Derwent Wood completed the more finished bust, which is now at Trinity. The Thomson family were surprised to learn of its existence when they saw it at the Academy Summer Exhibition. It was bought by Mr W.W. Rouse Ball, and presented to the College by him.

We may here give a list of the various portraits of J.J. Thomson.

Artist	Date	Location
Arthur Hacker, R.A.	1903	Cavendish Laboratory*
Fiddes Watt	1922	Royal Society
René de L'Hôpital	1923	Royal Institution
William Nicholson	1924	Trinity College Dining Hall

Drawings		
William Strang, R.A.	1909	Royal Library, Windsor Castle (O.M. Series)
Francis Dodd, R.A.	1920	Fitzwilliam Museum, Cambridge
Walter Monnington, A.R.A.	1932	National Portrait Gallery
Henrik Lund	1932	Oslo
Dr Foxley Norris (Dean of Westminster)	1935	(Replica given to Lady Thomson)

The drawing by Dr Foxley Norris was done secretly on the back of a menu during the Trinity Commemoration Dinner, and he was only detected by the ladies in the Master's Lodge who were

* See p. 145 above.

listening to the speeches from the open panel above the dais. He confessed when taxed with it, and gave Lady Thomson a replica of the drawing afterwards.

A 'talking film portrait' of J.J. Thomson was made in 1934 and is in the keeping of the Institution of Electrical Engineers, so that future generations should be able to form a notion of his personality as seen and heard by his contemporaries.

In earlier years he went with Lady Thomson to the Private View of the Academy, but later he could not find time for this. He went regularly to the Academy banquet year by year, which he looked forward to. In 1920 he spoke there in response to the toast of 'Science'. He also went with his daughter to the private views of several of the Winter Exhibitions. He was entirely delighted by the Dutch Exhibition of 1929 and said to one or two friends whom he met there: 'These old Dutchmen really did get the sunlight into their pictures.' The painting he liked best in this exhibition was Vermeer's view of Delft. He went also to the private view of the Persian, Italian and English Exhibitions. He did not care much for the Italian, and criticised Botticelli's 'Birth of Venus', saying 'it was ridiculous to paint a young woman who looked as if she weighed quite ten stone standing on the edge of a cockle shell, that would certainly tip over with her weight'. Leonardo's drawing of the Virgin and St Anne, which belongs to the Academy, was what he admired most.

At the English Exhibition he was eager to discover Millais' 'Autumn Leaves' from the Manchester Gallery, which he said had been his favourite picture as a boy.

In the matter of entertainment J.J. did not care much for concerts. He remarked that he had too often been sent to escort guests to the Hallé Concerts in Manchester in his youth. However, he usually attended the May Week Concert at Cambridge, and showed good taste as to whether the music was well performed or not.

He had been enthusiastic about Gilbert and Sullivan's operas at the time when they came out, chiefly in the '80's and '90's, and prided himself that he had seen the first nights of many of them. He was always ready to go to revivals of these, though apart from this he did not often go to the theatre in his later years. He appreciated Miss Ruth Draper's entertainments.

J.J. Thomson's tastes in reading could be diagnosed to some extent from an examination of the books in his study shelves. Most of his books of course related to physical science. His collection of herbals was there, and there were books relating to wild flowers, of which those by the Rev. C.A. Johns were in constant use. There were others describing the flowers in various parts of England. There was a good sprinkling of rather light novels, in addition to detective stories. At intervals Waverley novels would appear on all the tables, as the set of these which J.J. had won as a prize was kept in the drawing room, but he liked to re-read them from time to time. He was fond of Dickens. It was his custom to read *The Cricket on the Hearth* through on Christmas Eve, just before going to bed. He also liked Thackeray, and perhaps still more Trollope, and Jane Austen. The Brontës he considered were over-rated, but he admired Mrs Oliphant, though he said her poverty and the needs of her large family compelled her to over-write herself. George Eliot he appreciated, though he said she suffered from living too much with clever people. He particularly liked the humour of *Scenes from Clerical Life*. He saw the absurdities of Ouida's novels when these were in vogue, but he also recognised their merits. At a later date he was delighted with Mr J.B. Priestley's *Good Companions*, but not equally pleased with this author's later books. At all times he was fond of reading 'thrillers'—and his shelves were well stocked with them.

To MRS H.F. REID:

Trinity Lodge.   Dec. 21, 1930.

Very many thanks for 'Charlie Chan carries on', it was my first introduction to him and he gave me plenty of excitement. I don't think he observes all the rules of the game which ought to govern detective stories, I think if one does not guess the answer the first time of reading, on re-reading you ought to find some place where your suspicions should have been aroused, but I did not find one in Charlie.

I am sending a little book on History, which contains only two dates, one of which, 1066, I knew. It is not very good, but at present there is a great craze for people to read what they do not understand and recalls to my mind the Browning craze of 50 or more years ago, only now it is not poetry they read but scientific speculations by Jeans and Eddington; if one is supposed to know some physics the questions one gets asked by literary ladies are most embarrassing, you cannot very well tell them they have not understood one word of what they have been reading.

J.J. in fact, though he considered it helpful for a physicist to have to explain his ideas to untrained people at the right time and in the right place, disliked extremely having to attempt this in dinner-table conversation, or, generally, being asked scientific questions by laymen.

However, to return to his literary tastes. In his earlier days he might often be seen towards the close of the day at the Union, selecting novels from the library, and carrying them home under his arm.

Several shelves in his study were devoted to biographies. I have heard him speak with appreciation of Lockhart's *Life of Sir Walter Scott*, and Monk's *Life of Bentley*, which was so intimately related to the history of the Lodge. He was fond too of browsing in Boswell's *Johnson*.

Although as we have seen Thomson was prepared to admit in principle the claims of literary and philosophical studies, he did not often make any detailed allusion to history or philosophy. It is not remembered that the collection of books in his study suggested that he took any very personal interest in these subjects. In the course of compiling an obituary notice I had been writing an account of A.J. Balfour's philosophical views, and attempted to elicit what J.J. thought of them, but failed to get any reaction, though he knew Balfour well, and might have been expected to take some interest. He is remembered to have asked in the spirit of one willing to learn, whether philosophy had made any substantial advance since the time of Plato.

In pre-war years J.J. often went for his holidays to North Wales, together with Lady Thomson and the Battys. Frederick

Thomson had his Camp for the Boys' Brigade close by. J.J. spent his time in walking many hours a day and reading. Later he went to Shaftesbury, to the New Forest, to various places in Scotland, and to Hunstanton, where he played golf and searched for wild flowers.

Most of his visiting was done on the occasions when he went to a place for some definite function and received hospitality. Apart from that he visited a few friends who happened to possess country houses—Sir Charles Parsons, Lord Haldane, Prof. G.M. Trevelyan, and myself. Except for his visits to America and Canada he did not indulge much in foreign travel. He went to Paris to receive a degree at the Sorbonne in 1923 and again in 1927 to receive the Mascart Medal. Some of the French ladies were very interested to hear that J.J.'s first special attention to Lady Thomson during courtship was the loan of vol. 1 of Mascart's *L'Electricité et le Magnétisme*.

After about 1929 Lady Thomson, whose health had not for some time been very firm, ceased to be able to leave Cambridge, and J.J. himself was no longer able to take a great deal of exercise. He became much more willing to travel about.

In 1933 he went with his daughter Joan for an Atlantic cruise. At Madeira he was disappointed to find that tobogganing down the hill was less sensational than he had been led to expect. He insisted on going about in a bullock cart rather than a taxi. In the last two summer holidays he went for motor tours with his daughter. He was delighted when the hotel waiter at Stratford-on-Avon after helping Miss Thomson to what remained on the dish said, 'I will bring some more for your husband, madam'.

J.J.'s membership of *The Family* forms an interesting link with the past. This was, and is, a Cambridge dining club consisting of twelve members, each of whom acts as host once a year and each host tries to produce something unusual or exquisite in cooking. J.J. never missed a meeting during the fifty years which are within Lady Thomson's knowledge, except during the last few months of his life. Sir George Paget, J.J.'s father-in-law, had been a member, also for some fifty years, and his membership overlapped J.J.'s so that they divided a whole century between them. The Club had originated as a Jacobite Society, and one of Sir George Paget's fellow-members had been born before the rebellion of 1745.

J.J. was very fond of his garden. He describes how he had a small garden as a boy, and spent most of his pocket money on it. Once in an after-dinner speech he said that much of his leisure was spent in killing snails. He was apparently not quite happy about killing them because he spoke of the difficulty of deciding whether it was better to kill them in a pail with salt, or to throw them over the wall into the next garden. However, those who had the best opportunity of observing do not seem to know much about the snails, and they were perhaps apocryphal. He very occasionally did some watering, but little or nothing else in the way of manual labour. But for all that a great part of his happiness was derived from his garden. Almost every day he walked carefully round it, either alone or with his daughter, noticing what individual plants were doing. He took a great deal of trouble in choosing bulbs every autumn; many nurserymen's catalogues were accumulated in his study, and he even put off his final decision to accept the invitation of the Franklin Institute to go to America in 1923 until the bulb order had been sent off.

After he went to the Master's Lodge in Trinity he specialised in irises, which did very well in the garden there. Several of them were varieties raised by Sir Michael Foster,* and called after the wives of Foster's friends. J.J. was also very fond of rock plants. Part of the greenhouse was unheated so as to grow a collection of such in pots. These he chose personally from stalls in the Cambridge market place, and from nurserymen. He was very fond of the little Alpine house in the Cambridge University Botanical Garden, and became quite excited over the flowering of the little *Bulbocodium* and *Cyclameneus narcissi*. He had a friendly rivalry with his friend, J. D. Duff, as to who could get the best results with gentians.

He did not care for large showy flowers. Top-heavy dahlias and chrysanthemums did not appeal to him at all. Lady Rayleigh— the wife of the present writer—had some very special rhododen-

---

* Professor of Physiology at Cambridge, 1883–1903. An enthusiastic hybridiser.

drons at Beaufront Castle, near Hexham, but J.J. during several days' visit there did not take any notice of them, or apparently even discover that a common interest existed. In his own home he has been heard to abuse a large rose in a vase on the dinner table, saying it was 'blowsy' and that there was nothing to compare with the English wild rose. Good colour was the thing that appealed to him most of all in flowers, and he seemed to have a real affection for little plants.

He was just as fond of wild flowers as of garden ones. His great holiday interest was to look for the characteristic wild flowers of whatever place he happened to be in. Regularly every year he went to Hunstanton for a few days' holiday at the end of June. He liked the place, which suited his health; he played golf there, and after he had given that up, he still went and looked for the wild flowers to be found on the links. He was chairman of the Wicken Fen Committee of the National Trust, formed to deal with a bequest of Charles Rothschild, a son of the first Lord Rothschild, who was a Trinity man and an enthusiastic naturalist. The testator's object was to preserve this piece of unspoilt nature for the nation. J.J. thoroughly enjoyed the annual excursion to the Fen, and used to talk of it with interest and pleasure.

One of the subjects which he emphasised when addressing young people at prize givings and the like was the pleasure to be got from scenery, even when it is nothing out of the common. The best things, he said, were not as a rule those chosen for picture postcards. Boys should learn to appreciate for example the beauty of a field of corn swayed by the wind, and they might gain much happiness from such things.

He liked to go every year to see the *Anemone pulsatilla* in one of the two places near Cambridge where it is found. He never grudged time or trouble in looking up the names of wild flowers, which he did not already know, whether found by himself or other people.

From SIR W.B. HARDY:

5 *Grange Road, Cambridge.*   17. VII. 1928.

In or about a pond on Quy Fen, last Sunday,

1. Common Bladderwort in full flower.

2. *Alisma ranunculoides* in flower; a handsome plant.
3. Bog pimpernel *in profusion*.
4. Sneezewort (*Achillea Ptarmica*).

I was surprised by (4) which we had found previously only in Cornwall. (1) is an interesting plant.

Did you find Man Orchis this year?

J.J. never seemed too busy to listen to conversations on wild flowers—even those of countries which he had not himself visited. He collected herbals, which he bought from second-hand book-sellers in Cambridge. After his death, these were found to be worth more than he had paid for them.

In his old age he used to say that if he were to begin his scientific life over again he would become a botanist. Notwithstanding his fondness for the naming of wild flowers, he did not mean that he would be a systematist, but rather a plant physiologist. I think he pictured himself as having some success in unravelling the mysteries of growth and the distinction between the living and non-living. Whether he *would* have succeeded in this distinction it is impossible to guess. We do not know how far off the goal may be.

The following, which I have from Prof. F.L. Hopwood, is of interest in this connection. At a dinner given in 1927 to J.J. by the Röntgen Society, Dr G.W.C. Kaye turned to him and asked: 'I suppose, Sir, if you were choosing a career to-day you would still choose to be a physicist?'

'No,' replied J.J. 'I should become a biologist; I consider that biology is now in the same stage of advancement that physics was when I took it up forty years ago.'

The Master of Trinity was often observed by undergraduates looking into shop windows. They thought of him as the absent-minded professor whose eyes might seem to be looking at mundane objects, but whose mind was far away, fixed on some abstruse line of thought. In this they were entirely wrong. He was taking careful note of the things displayed, and could often tell his family where (e.g.) good dessert fruit could be bought. He was very fond of going into Woolworth's to look round. When someone remarked that it was uncomfortably crowded he replied: 'Not if you know the right time to go.' It would seem that his interest was really in the business methods which he saw in practice there. His purchases usually took the form of stationery, and he accumulated a store of shiny red-backed exercise books, intended for children, with 'Rule of the Road' instructions on the backs. He also bought detective stories and occasionally simple sweets such as peppermints. He went to other shops as well, and brought home small alpine plants from the market and cigarettes. He was also fond of going to Boots' and often praised the firm's methods, saying how well the employees were treated. He took an interest in the Cambridge antique shops, and had his own opinion as to which of them were reliable.

On the other hand he was not equally interested in tailors and outfitters, and diplomacy had to be used to get him to make the necessary visits to them. Although he took no interest in his own dress, he was by no means equally indifferent to the dress of the ladies of his family. He was always willing to help in the choice of patterns. He made surprisingly observant comments on the dress of ladies who came to the house, and always seemed able to detect whether a dress was well made or not.

His interest in shops was most marked when he was away for a holiday, and he generally chose a present to take back to Lady Thomson. A village store would fascinate him, he would study the window and discuss with the shopkeeper what business was done and the prices of goods.

His outlook was in many respects individualistic. He liked private enterprise and often quoted the success of Morris (Lord Nuffield) the motor car manufacturer in favour of the private bank, saying that Morris was lent the money which enabled him to make his first start by a local banker who trusted him personally.

We have seen that the celebration of J.J.'s 70th birthday was in the main a Cavendish Laboratory festival. The chief activities of his later years had been transferred to Trinity, and it was fitting that his 80th birthday (December 18th, 1936) was celebrated there. Letters of congratulation came from many quarters, at home and abroad. One birthday present is of special interest. It was a manuscript of one of Heinrich Hertz's scientific papers, destined for the library at Trinity, and was sent by Frau Hertz, who was resident at Cambridge. With it was a letter expressing warm thanks for Thomson's kindness to her, in having done much to make it possible for her to take refuge in Cambridge with her daughters. There were also presents of flowers from various quarters. On the morning of his birthday, a deputation, headed by Lord Rutherford, called at the Lodge and presented a bound address, signed by workers in the Cavendish Laboratory, and by other scientific people in the University.

From SIR WILLIAM BRAGG (after conveying the congratulations of the Royal Institution):

December 17th, 1936.

I must be allowed to add my own personal congratulations. Just fifty-one years ago, I was walking with you along the K.P. on our way to the Cavendish where you were going to lecture and I was going to be one of the audience. You asked a chance question, which sent me off to the telegraph office after the lecture was over and I applied for the Adelaide post which Lamb was vacating. It was the last day of entry; and of course your remark sent me to Australia. Perhaps you were the one who asked a certain Adelaide man—then visiting London—whether the Council of the University of Adelaide was likely to prefer a senior Wrangler who occasionally disappeared under the table after dinner to a young man who so far had shown no signs of indulging in the same way. The Adelaide man was Sir Charles Todd, whose daughter I married a few years afterwards.

You see I have a special reason for writing to you on an occasion like this.

The actual celebration took the form of an address and a present from the Fellows. Owing to the season, the number present was not too large for the Combination Room, and the dinner was held there informally, morning dress being worn. Fifty-three of the Fellows were present. The presentation was made by the Vice-Master, Mr D. A. Winstanley. In the course of his speech he said:

As it is difficult to please all men, it is possible that some of your eminent predecessors would not have commended you as highly as we do. Bentley would have disliked the cordiality of your relations with the Fellows, and Whewell would have disapproved of your friendly interest in the undergraduates and their sports and pastimes. But the present Society does not subscribe to the opinions of these distinguished men, and although in a volume of reminiscences recently published it is truly remarked that Fellows of Trinity are slow to express sentiments of approval there are occasions when they are not slow, and this is one of them.

The presentation consisted of a copy of the Boyle Cup, made by John Bodington in 1697 and given to the College by Henry Boyle. With it was an address, which ran as follows:

The Ancients who first fabled that the world was made of atoms taught that these wayward bodies move, for no discoverable reason, in any and every direction. You, Master, to whom the habits of the atom are as a book unsealed, must have remarked how nearly they resemble the behaviour of the members of our Society. When we are called upon to determine whether the fountain shall be bordered with flowers or the chapel beleaguered with iron spikes, we meet only to rebound and the predictable outcome of a collision is that we shall be found inextricably revolving in a vortex.

There are, however, some few occasions when the unseen harmony which enables us to differ passionately on the most trivial matters with no loss of mutual esteem, becomes happily apparent, an amicable impulse of gravitation sets us moving towards a common goal. On your eightieth birthday, Master, we unite in offering a token of affectionate regard. We recall with pride that you were a leader among those pioneers whose imagination has opened an immeasurable vista beneath the bounds of human sight and has read in the dust of earth the secrets of the stars. Your name will stand with those few who have enhanced the peculiar fame bequeathed to our Society by Newton. No less than by your achievements in the pursuit of learning, you have deserved well of the college by your unfailing interest in every one of its activities, and by your patient and impartial kindliness in presiding over our deliberations.

Together with your family and other friends we wish that your days may be prolonged in the enjoyment of goodwill and honour, with peace of mind.

J.J. was visibly affected, and spoke as follows:*

It is difficult to express by words deep emotion—the author of the saying 'out of the fullness of the heart the mouth speaketh' must have been a great orator, which I most certainly am not, or had never made a speech when his heart was full, which is what I am trying to do now. What I rely upon is the fact that this is a domestic occasion, and that I am speaking to those who have had so much practice in pardoning my imperfections that I know they feel that failure on my part to express my feelings is not because I do not feel but because I lack the ability to express them.

Some of those that are here this evening have spent much time over the entrance scholarship examinations, and that has taken my mind back for just over sixty years when I was a candidate, for there was one incident in it which I remember almost as vividly as when it occurred. I had formed in my own mind a picture of a college tutor. At the Owens College where I came from there were only professors, who were not concerned with discipline. I knew that college tutors, besides lecturing, were responsible for discipline and I pictured to myself a tutor as a man with the learning of a professor and the austerity of a schoolmaster. When I gave up the first paper I took I found myself confronted with a little man in spectacles looking very much like what I had imagined Mr Pickwick to look without a trace of austerity about him. I asked some of the other candidates who he was, and one of them said his name was Prior.† I was so surprised at his mild and kindly aspect that it fixed itself vividly on my memory, and I can recall the scene after sixty years. I remember the way he took up the paper, looked at it through a corner of his eye to see I had written my name on it before he put it into the basket.

I got a minor scholarship, and it meant everything to me. I cannot believe that if I had gone into anything different I should have had nearly so happy a life as I have had at Trinity. I cannot say too much of the kindness I have received from you all. When I was appointed Master I was haunted by the fear that the College would have much preferred someone else, since up till then I had taken no part in the

---

* The speech is edited from imperfect notes. I have used the ordinary licence of a reporter, but have added nothing to the sense. The notes break off in the middle of a sentence.

† Joseph Prior. Twelfth Wrangler in 1858. Tutor of Trinity 1870 to 1886. Died 1918. His knowledge of mathematics was generally considered to be somewhat limited, but socially he was a most amusing companion, and a feature of the high table.

Administration. [Here the notes break off. The speech concluded with the words] God bless you all.

It seemed to those present that it came somewhat as a revelation to J.J. on this occasion how warm were the feelings of the Fellows generally towards him. It was most fortunate that he was able to take part in this function while he was still his old self. The speeches were good and it was a successful evening.

However, the shadows were now visibly lengthening and this seems to be an appropriate place in which to collect what little there is on record about J.J.'s private religious views and devotions. He was, in general, reserved about this like most of his fellow-countrymen, though he made it clear that he respected the sincere views of others even in such unpopular directions as conscientious objection to military service. On the other hand he made no secret of his dislike of Anglo-catholicism—though not necessarily of Anglo-catholics personally. On the more fundamental questions he did not often expand, even to his family; but he did say on one occasion that he regarded Christ's death as the outcome of His mission to save the world by bringing to it an ideal of unselfishness and aspiration to heavenly things. The sacrifice of His own life resulted.

At the end of a discussion about the modern Christian teaching with a young Fellow of Trinity who had gone as a missionary to India, he said that he had great sympathy with Pontius Pilate, and his question, What is truth?

The following correspondence with Canon Edward Lyttelton* throws some light on his religious attitude. Canon Lyttelton wrote from Overstrand, near Cromer, March 15th, 1934.

When we had a talk at the Pepys dinner at Magdalene you did me an honour by asking some questions about the High Church teaching and position and the warrant alleged for it in the teaching of Christ. I felt at the time dissatisfied with the answer I tried to give; and that it was not fitting that you should be fobbed off with anything but the best that I could give. So I thought you wouldn't mind if I set down on paper a few thoughts which have come to me since early days at Cambridge. The only claim I make for them is that they are not simply

* Canon of Norwich. Headmaster of Eton, 1905–16.

echoes of other men's teaching—however eminent—but corollaries drawn from deep conviction which I am thankful to know are held in common by us both. There is a certain compensation for ignorance; that one is led to think at first hand, guided but not dragged along by men who know.

Canon Lyttelton's views are given at too great length for insertion in full, but the following seems to be a fair summary of them.

Anyone who continues through life in the practice of private prayer and in recognition of the Sacraments must assume that the Creator *acts*. Can we discern any law by which God acts? His gifts of intellect, wealth, physique, and the like are given to the few, and this must be in order that they may be used for the benefit of the multitude. It is *a priori* more than likely that He would choose a group to whom He would entrust the precious knowledge for distribution. For such a group to be effective they must act corporately, and be organised. This organisation, Canon Lyttelton sees in the Church, in spite of its failings in the internecine strife and in other ways. It is in fact in this way that the Gospel has been preserved from corruption, and caused to spread to the great advance of civilisation. No words of Christ can be found laying this down. It was not His method. He left it to his disciples, merely promising that He would be with them.

J.J. Thomson wrote in reply, from Trinity Lodge:

March 31st, 1934.

I am very grateful to you for your letter. I have read it twice and find it very clear and extraordinarily interesting. I am in sympathy with all of it, and in agreement with nearly all, and look forward to having another talk with you on these matters.

It may be added that J.J. himself was one of those who 'continues through life in the practice of private prayer and in recognition of the Sacraments'. He was a regular communicant, and it was his invariable practice to kneel in private daily prayer. He continued to do so till the end of his life, even under circumstances which made it physically very difficult. He is remembered by his daughter to have spoken of the very great benefit derived from the effort of prayer. 'That being so,' he said, 'one naturally turns to it.' He usually retired to bed very late, 1 a.m. or 1.30, or even later. He always paused at the bookcase near the door wherein his small Bible stood, and spent two or three minutes in reading it. Until near the end of his life, his habits of private devotion were only known to Lady Thomson.

It remains to say a few words about the closing years. It seemed to some of those who knew him best that in his old age Thomson's mental activity was most conspicuous as a man of affairs and as a mathematician. As a physicist his intellectual vigour, and perhaps also his interest, seemed to have sensibly declined. When assessing others he seemed to lay more stress than of old on intellectual thoroughness and grasp, and less than of old on original achievement.

His facility as a mathematician never deserted him. Sometimes in his old age he saw the papers set in the mathematical tripos, and amused himself with trying whether his skill was still equal to dealing with them; and he did not often fail.

I remember seeing him leaving St Paul's clad in a scarlet Doctor's gown on the occasion of King George V's Silver Jubilee in 1935, and he visited London after that, though crossing the streets was rather an anxiety. In May 1938 he went up to receive the Kelvin Medal. I was commissioned by the professional Engineering Institutions to make the presentation, and I endeavoured to explain the bearings of his work on Engineering, but he protested in his reply that, notwithstanding what I had said, as an engineer he was a complete fraud!

Miss Gertrude Mellor writes:

The last time I was at Cambridge [probably in the summer of 1938] I saw him running by the side of the river during the Races Week to be in at the Bumps at the finale of the races.

He continued to attend the annual Cavendish Laboratory dinners except the last one, only a few months before the end.

In September 1938 the British Association met at Cambridge, and the Thomsons entertained a party at the Lodge. J.J. appeared, probably for the last time, in the lecture room of the

Cavendish Laboratory, and 'read a paper' or rather gave an exposition to Section A which met there. He failed to make his points very clear, and I could not help feeling saddened at the contrast with what I had heard from him in that same room forty years before.

During the Munich crisis (September 1938) he wrote of Neville Chamberlain: 'The Prime Minister is magnificent.' How many who said this have now forgotten their own words? There was in my view (and I had the opportunity of observing both at close quarters) a certain affinity of character between J.J. and Neville Chamberlain in the simple and straightforward point of view taken by each, and in their grasp of what was within the bounds of practicability, in their readiness for reasonable compromise and their willingness to exchange ideas with anyone who had something to say.

At the annual Trinity Commemoration of Benefactors in March 1939 he presided as usual, but as he complained in the course of the speech, his memory was failing him; and he was not in fact able to hold the attention of his hearers. He was, however, quite himself in some conversation I had with him the next morning. I called and saw him for the last time on November 23rd, 1939, and was warned that he was not well and should not be pressed on any subject: however, there was nothing noticeably wrong, except his laboured breathing as he shuffled slowly along the passage. He seemed pleased to see me and took a sanguine view about the war, which was then in a dormant stage. At this time he was doing little or nothing in the way of College administration, the Vice-Master acting for him. He continued to attend Hall on Sundays for the earlier part of 1940. His last public act was to read the lesson at the funeral of his great friend, J.D. Duff, in the College Chapel at the end of April 1940.

In the summer he did not leave the Lodge, except to sit in his garden. About six weeks before he died, when he was failing very much and hardly able to attend to most subjects, his daughter, who had been out riding, was telling him of the wild flowers she had seen, and failed to recall the name of one of them. But he at once gave it from her description.

One of the Fellows, who visited him two days before the end, tells that his last question was whether all was well with the College.

He died on August 30th, 1940.

The funeral (after cremation) was in Westminster Abbey near the graves of Newton, Darwin, Herschel, Kelvin and Rutherford in the nave. It was conducted by the Dean and the Abbey clergy. There was at the same time a Memorial Service in the Chapel of Trinity College, conducted by Dr H. F. Stewart and the Rev. B. Dennis Jones.

There would be no more fitting epitaph to J. J. Thomson than the reference to him made by his successor, Prof. G. M. Trevelyan, at the time of his inauguration.

I am incompetent to judge of the late Master's scientific achievement. But Professor Bragg has written of it: 'He, more than any other man, was responsible for the fundamental change in outlook which distinguishes the physics of this century from that of the last.'

I will leave it at that. But we here to-day remember 'J.J.' as our friend, his unaffected modesty, the most beautiful of all settings for superlative powers of mind; his ever-active love of the College; his interests in its athletic as well as its academic successes and failures from day to day; his evident desire to be regarded as an ordinary plain man among ordinary plain men, though his genius had in fact raised him so high above our heads.

APPENDIX I

LIST OF DISTINCTIONS

UNIVERSITIES

Minor Scholar of Trinity College, Cambridge	1876
Major Scholar of Trinity	1878
Sheepshanks Astronomical Exhibitioner	1879
Second Wrangler, B.A.	1880
Second Smith's Prizeman	1880
Fellow of Trinity College	1880
Adams Prize, Cambridge University	1880
Rede Lecturer, Cambridge University	1896
Master of Trinity College	1918
Honorary Doctor of Science, Cambridge University	1920
Honorary Doctor, Dublin University	1892
Honorary Doctor of Laws, Princeton University	1896
Honorary Doctor, Victoria University, Manchester	1900
Honorary Doctor of Philosophy, Cracow University	1900
Honorary Doctor, Glasgow University	1901
Honorary Doctor, Birmingham University	1901
Honorary Doctor of Laws, Johns Hopkins University, Baltimore	1903
Honorary Doctor of Science, Columbia University	1903
Honorary Doctor, Oxford University	1904
Honorary Doctor of Science, Berlin University	1905
Honorary Doctor, University of Göttingen	1905
Honorary Doctor of Laws, Aberdeen University	1906
Honorary Doctor of Science, Sheffield University	1910
Honorary Doctor of Laws, St Andrews University	1911
Honorary Doctor of Philosophy, Oslo University	1911
Honorary Doctor, Sorbonne	1923
Honorary Doctor, Philadelphia University	1923
Honorary Doctor of Laws, Edinburgh University	1930
Honorary Doctor of Science, London University	1931
Honorary Doctor, Leeds University	1933
Honorary Doctor, Reading University	1935
Honorary Doctor, Athens University	1937

GOVERNMENTAL

Knight Bachelor	1908
Order of Merit	1912

MEDALS, ETC.

Royal Medal, Royal Society	1894
Hughes Medal (1st time Awarded) } Royal Society Copley Medal	1902
Hodgkins Medal, Smithsonian Institute, Washington	1902
Nobel Prize for Physics	1906
Elliot Cresson Gold Medal, Franklin Institute, Philadelphia	1910
Albert Medal, Royal Society of Arts	1916
Franklin Medal } Philadelphia Scott Medal	1923
Faraday Medal, Society of Electrical Engineers	1925
Gunning Victoria Jubilee Prize, Royal Society of Edinburgh	1925
Mascart Medal, Paris	1927
Sylvanus Thompson Medal, Röntgen Society	1927
Dalton Medal, Manchester Literary and Philosophical Society	1931
Kelvin Medal, Institutions of Civil, Mechanical and Electrical Engineers	1938
Honorary Freedom of the Worshipful Company of Salters	1922
Honorary Freedom of the Worshipful Company of Grocers	1924

LEARNED SOCIETIES, ETC.

BRITISH

Fellow of the Royal Society	1884
Vice-President of the Royal Society	1911–1913
President of the Royal Society	1915–1920
Honorary Professor of Natural Philosophy, Royal Institution	1920
President of Section A, British Association (Liverpool)	1896
President of British Association (Winnipeg)	1909
President of Section A (London)	1931
President, Cambridge Philosophical Society	1894
Honorary Member, Manchester Literary and Philosophical Society	1895
Honorary Member, Royal Irish Academy	1900
Honorary Fellow, Royal Society of Edinburgh	1905
Honorary Member, Institution of Electrical Engineers	1907

Corresponding Member, Royal Society of Canada   1909
President, Junior Institution of Engineers   1910
Honorary Fellow, Optical Society   1910
President, Faraday Society   1911
President, Physical Society of London   1914
Honorary Member, Academy of Sciences of Bengal   1915
Honorary Member, Royal Society of New South Wales   1915
Honorary Member, Institute of Metals   1916
Honorary Member, Röntgen Society   1918
Honorary Fellow, Royal Society of Medicine   1919
Honorary Member, Royal Engineers Institute   1920
Honorary Member, Institute of Physics   1921
President, Institute of Physics   1921
Honorary Member, Royal Philosophical Society of Glasgow   1924
Honorary Member, Institution of Civil Engineers   1925
Honorary Fellow, Chemical Society   1927
President, Association of Special Libraries and Information
     Bureaux   1929
Honorary Member, Institution of Mechanical Engineers   1932

FOREIGN

Foreign Correspondent, Royal Academy of Sciences of Turin   1896
Member, Royal Society of Sciences of Upsala   1901
Member, Reale Accademia dei Lincei, Rome   1903
Honorary Fellow, National Academy of Sciences, United
     States of America   1903
Member, American Philosophical Society, Philadelphia   1903
Member, Physical Medical Society, Erlangen   1903
Honorary Member, Royal Academy of Sciences, Amsterdam   1904
Member, Bologna Academy of Sciences   1904
Corresponding Member, Royal Bavarian Academy of Sciences   1907
Foreign Member, Royal Swedish Academy of Sciences   1908
Member, Videns Kabs sels Kabel Christiania (Oslo)   1908
Member, Royal Academy of Sciences of Borussica   1910
Member, Royal Academy of Sciences, Göttingen   1911
Honorary Member, Imperial Society of Devotees of Natural
     Sciences etc., Moscow   1911
Correspondent, French Academy of Sciences   1911
Corresponding Member, Imperial Academy of Sciences,
     St Petersburg   1913

Member, French Academy of Sciences	1919
Member, Belgian Academy of Sciences	1919
Foreign Member, Società Italiana delle Scienze	1919
Member, Royal Danish Academy of Sciences	1920
Honorary Member, Franklin Institute, Philadelphia	1922
Honorary Member, Russian Academy of Sciences	1925
Member, Società Reale di Napoli Accademia delle Scienze Fisiche e Matematiche	
Foreign Member, Polish Academy of Science	

APPENDIX II

SUPPLEMENT TO BIBLIOGRAPHY

A full bibliography of Thomson's writings is given in the notice of him by the present writer, published in the Royal Society's Obituary Notices for 1941. It is not proposed to repeat it here, but the following items were overlooked:

1927. *Dictionary of National Biography.* Obituary notice of J.H. Poynting.

1910. *Encyclopaedia Britannica,* 11th Edition. Articles: 'Conduction of Electricity through Gases' and 'Electric Waves'.

1922. *Encyclopaedia Britannica,* 12th Edition. Article: 'Gases, Electrical Properties of'.

1926. *Encyclopaedia Britannica,* 13th Edition. Article: 'Gases, Electrical Properties of'.

1929. *Encyclopaedia Britannica,* 14th Edition. Article: 'Electric Waves and Electricity, Conduction of' (in part).

INDEX

Milton Keynes UK
Ingram Content Group UK Ltd.
UKHW041521181024
449640UK00009B/122

9 781107 655423